IET MANAGEMENT OF TECHNOLOGY SERIES 27

How to Build Successful Business Relationships

Other volumes in this series:

Volume 15	**Forecasting for technologists and engineers: a practical guide for better decisions** B.C. Twiss
Volume 17	**How to communicate in business** D.J. Silk
Volume 18	**Designing businesses: how to develop and lead a high technology company** G. Young
Volume 19	**Continuing professional development: a practical approach** J. Lorriman
Volume 20	**Skills development for engineers: innovative model for advanced learning in the workplace** K.L. Hoag
Volume 21	**Developing effective engineering leadership** R.E. Morrison and C.W. Ericsson
Volume 22	**Intellectual property rights for engineers, 2nd edition** V. Irish
Volume 23	**Demystifying marketing: a guide to the fundamentals for engineers** P. Forsyth
Volume 24	**The art of successful business communication** P. Forsyth

How to Build Successful Business Relationships

Frances Kay

The Institution of Engineering and Technology

Published by The Institution of Engineering and Technology, London, United Kingdom

© 2009 The Institution of Engineering and Technology

First published 2009

This publication is copyright under the Berne Convention and the Universal Copyright Convention. All rights reserved. Apart from any fair dealing for the purposes of research or private study, or criticism or review, as permitted under the Copyright, Designs and Patents Act, 1988, this publication may be reproduced, stored or transmitted, in any form or by any means, only with the prior permission in writing of the publishers, or in the case of reprographic reproduction in accordance with the terms of licences issued by the Copyright Licensing Agency. Inquiries concerning reproduction outside those terms should be sent to the publishers at the undermentioned address:

The Institution of Engineering and Technology
Michael Faraday House
Six Hills Way, Stevenage
Herts, SG1 2AY, United Kingdom

www.theiet.org

While the author and the publishers believe that the information and guidance given in this work are correct, all parties must rely upon their own skill and judgement when making use of them. Neither the author nor the publishers assume any liability to anyone for any loss or damage caused by any error or omission in the work, whether such error or omission is the result of negligence or any other cause. Any and all such liability is disclaimed.

The moral rights of the author to be identified as author of this work have been asserted by her in accordance with the Copyright, Designs and Patents Act 1988.

British Library Cataloguing in Publication Data
A catalogue record for this product is available from the British Library

ISBN 978-0-86341-956-0

Typeset in India by Newgen Imaging Systems (P) Ltd, Chennai
Printed in the UK by Athenaeum Press Ltd, Gateshead, Tyne & Wear

Contents

Preface		**The nature of networking**	**xiii**
	1	How to progress your career and acquire valuable contacts and strong allies along the way	xiv
	2	So how do you go about making the right connections for yourself and your company?	xiv
1		**What is meant by business relationships?**	**1**
	1.1	Introduction	1
	1.2	A bit of self-management	1
	1.3	Making the right impression	2
	1.4	Prioritizing	5
	1.5	Self-management and being effective	6
	1.6	Getting the best out of feedback	6
	1.7	Seeing the bigger picture	8
	1.8	How do you come across?	10
	1.9	Managing other people's objections	11
	1.10	Awareness of others and flexibility	12
2		**Why do people network?**	**15**
	2.1	R for relationships	15
	2.2	Making a start	16
		2.2.1 Business introductions	16
		2.2.2 Strategic alliances	16
		2.2.3 Client relationships	17
		2.2.4 Team building	17
		2.2.5 Succession planning	17
	2.3	The three-stage plan: how it works	17
		2.3.1 The importance of networking	18
		2.3.2 Developing sucessful working relationships	25
		2.3.3 Anger management	29
3		**Making connections**	**33**
	3.1	Attitudes and approaches	34
		3.1.1 Responsiveness	34
		3.1.2 Review – a quarterly or half-yearly update	35

		3.1.3	Check – who really is who?	35
		3.1.4	A systematic approach	35
	3.2	Applied thinking		36
	3.3	Categorizing your business connections		36
	3.4	Suggested methods of connecting		37
	3.5	Coincidences do occur		37
	3.6	New approaches		38
	3.7	The case for developing business connections		39
	3.8	Reasons to get started		39
	3.9	Planning the meeting		40
	3.10	Go on a charm offensive		41
	3.11	Mastering the art of good questions		42
	3.12	Seven steps for making connections		42
		3.12.1	Step 1: Database	42
		3.12.2	Step 2: Categories	43
		3.12.3	Step 3: Links	43
		3.12.4	Step 4: Keeping in touch	43
		3.12.5	Step 5: Keeping notes	43
		3.12.6	Step 6: Being distinctive	43
		3.12.7	Step 7: Proactivity	44
	3.13	Relationship building		44
		3.13.1	A two-way process	44
		3.13.2	Watch your step	45
		3.13.3	Keep on track	46
		3.13.4	A learning curve	46
		3.13.5	Set objectives	47
		3.13.6	Make it a habit	47
		3.13.7	Creating a method	48
		3.13.8	Centres of influence	48
		3.13.9	Chance contacts	48
		3.13.10	Expect the unexpected	49
		3.13.11	When is the best time?	49
		3.13.12	What to say	49
		3.13.13	Persistence pays	50
		3.13.14	Say it with feeling	50
		3.13.15	Touching emotions and intellect	51
4	**Where to make a start**			**53**
	4.1	The ins and outs of internal workplace relationships		53
	4.2	Colourful characters		54
	4.3	Attitude counts		54
	4.4	Understanding the characters		56
		4.4.1	Destructive characters	56
		4.4.2	Disgruntled characters	56
		4.4.3	Deadpan characters	57

	4.5	Weapons of mass destruction		57
	4.6	How difficult are you?		58
		4.6.1	Self-knowledge helps	58
		4.6.2	Influences	59
		4.6.3	Control mechanisms	59
		4.6.4	The perfect solution	59
	4.7	Office politics and workplace relationships		60
		4.7.1	Intentions	61
		4.7.2	Appearances can be deceptive	62
		4.7.3	Keeping a watchful eye	62
		4.7.4	Best foot forward	63
	4.8	Meetings and how to conduct them		64
		4.8.1	Meeting fatigue	64
		4.8.2	A communication tool	64
		4.8.3	Effective meetings	65
5	**Rising to new challenges**			**69**
	5.1	Three aids to a networking strategy		71
		5.1.1	Aid 1: Confidence	71
		5.1.2	Aid 2: Knowledge	71
		5.1.3	Aid 3: Focus on the positive	72
	5.2	Focus on priorities		72
	5.3	A good beginning		73
	5.4	More reconnaissance		74
	5.5	Taking the plunge		74
	5.6	Mix 'n' match		76
	5.7	How to win friends and influence people		76
	5.8	Identifying key players		78
		5.8.1	Influential people	78
		5.8.2	Movers and shakers	79
		5.8.3	Corporate citizens	80
	5.9	What colour are you?		80
	5.10	Watch the body language		81
		5.10.1	Eyes right	81
		5.10.2	Posture	81
		5.10.3	Heads up	81
		5.10.4	Disarming yourself	82
		5.10.5	Legs	82
		5.10.6	Hands	82
		5.10.7	Distance	82
		5.10.8	Ears	82
		5.10.9	Mouth	82
	5.11	Finding the right opportunities		83
	5.12	Some strategic advice		83
	5.13	Creating a virtual team		83

6	**How to get (the communication) going**		**85**
	6.1	What happens when you open your mouth?	85
	6.2	Using your voice effectively	86
		6.2.1 Rhythm	86
		6.2.2 Speed	87
		6.2.3 Volume	87
		6.2.4 Pitch	87
		6.2.5 Pause	87
		6.2.6 Projection	87
	6.3	Reviewing your vocal skills	88
	6.4	Face-to-face encounters	88
		6.4.1 Listen carefully	88
		6.4.2 Be disciplined	88
		6.4.3 Be reflective	89
		6.4.4 Pitfalls to avoid	89
	6.5	The five levels of listening skills	89
		6.5.1 Level 1	89
		6.5.2 Level 2	89
		6.5.3 Level 3	89
		6.5.4 Level 4	89
		6.5.5 Level 5	90
	6.6	Directing the communication cycle	90
	6.7	Opening rituals	91
	6.8	Good conversational techniques	91
		6.8.1 Things to look out for	92
	6.9	Other forms of communication	92
		6.9.1 Telephone calls	92
		6.9.2 Voicemail messages	92
		6.9.3 Text messages	93
		6.9.4 Written communication	93
		6.9.5 Letters	93
		6.9.6 Email	93
	6.10	Communication-skills awareness checklist	93
	6.11	Manners and mannerisms	94
	6.12	Non-verbal communication	94
	6.13	Maintaining good morale	95
	6.14	Dealing with professional envy	97
		6.14.1 Tackling the situation	98
	6.15	Communicating with confidence	98
	6.16	Visual impressions	99
		6.16.1 Looking the part	99
		6.16.2 Eye contact again	99
	6.17	Good manners	100
	6.18	Pay attention	100
	6.19	Switching off the mobile	101
	6.20	Confidence boosting	101

7	**Managing other people**		**103**
7.1	Delegation		103
	7.1.1	Some reasons not to delegate	104
	7.1.2	The skill of delegating	104
	7.1.3	What should be delegated	105
	7.1.4	What should not be delegated	105
	7.1.5	How to delegate	106
	7.1.6	Advantages to project team members	106
	7.1.7	Delegation exercise	106
	7.1.8	A good delegator or a willing martyr?	107
	7.1.9	Key action points to remember on delegation skills	109
7.2	Appraisals		109
	7.2.1	Performance appraisals	109
	7.2.2	Benefits to the individual	109
	7.2.3	Benefits to the line manager	110
	7.2.4	Benefits to the client	110
	7.2.5	Preparation for the discussion	110
	7.2.6	Conducting the interview	110
	7.2.7	Active listening	111
	7.2.8	Constructive feedback	111
	7.2.9	Appraisee's Charter	111
	7.2.10	Appraiser's Charter	112
7.3	Disciplining staff and problematic colleagues		112
	7.3.1	Setting a good example	112
	7.3.2	Bulling and harassment	113
	7.3.3	Discipline versus punishment	113
	7.3.4	Possible solutions	113
	7.3.5	Summary/check list	114
7.4	Recruiting and selecting the right people		115
	7.4.1	Defining the job to be done	115
	7.4.2	Specifying the profile of the likely candidate	115
	7.4.3	Sources of candidates and methods of attracting the right person	116
	7.4.4	Assessing written personal details	117
	7.4.5	Systematic approach to interview	118
	7.4.6	Assessment and checks	119
	7.4.7	Final selection and appointment	120

8	**Is it working OK?**		**121**
8.1	Meeting the team		121
8.2	Working together		122
8.3	Team dynamics		123
8.4	Self-sufficiency and group dynamics		123
	8.4.1	Involvement	123
	8.4.2	Empowerment	123
8.5	The power of responsibility		124

	8.6	Strengthening the team		124
	8.7	Motivation		124
		8.7.1	A brief overview	124
		8.7.2	Motivating the team	125
		8.7.3	Obtaining significant motivational results	126
		8.7.4	Getting the best from the team	126
		8.7.5	Consistency counts	128
		8.7.6	Why motivation matters	128
		8.7.7	The fundamentals of motivation	129
		8.7.8	Theory X and Theory Y	130
		8.7.9	Herzberg's motivator/hygiene factors	130
		8.7.10	Producing positive results	132
		8.7.11	Involving people	134
		8.7.12	Aiming for excellence	137
9	**Results, referrals, rewards**			**139**
	9.1	Keep up the numbers		139
		9.1.1	Rule 1: Create empathy	140
		9.1.2	Rule 2: Be courteous	140
		9.1.3	Rule 3: Be enquiring	140
		9.1.4	Rule 4: Create interest	140
		9.1.5	Rule 5: Show respect	141
	9.2	Make it real		141
		9.2.1	Sincerity	141
		9.2.2	Speak plainly, don't use jargon	141
		9.2.3	Why should people respond to you?	141
		9.2.4	Brand and company identity	142
	9.3	Recommendations and referrals		142
		9.3.1	Pareto's Law	143
		9.3.2	Why recommendations and referrals work	143
	9.4	Every relationship is different		144
	9.5	Advanced relationship-building skills		145
		9.5.1	What defines 'persuasive'	145
		9.5.2	Make what you say attractive	146
		9.5.3	Talk about the benefits	146
		9.5.4	Make it credible	146
		9.5.5	Add value	147
		9.5.6	The fun factor	147
		9.5.7	The personal touch	148
		9.5.8	Hold on to your assets	149
		9.5.9	Realize their potential	150
		9.5.10	Giving recognition for good service	150
		9.5.11	External relationships	150
	9.6	Further ways to develop professional relationships		154
		9.6.1	Competitor Analysis	154
		9.6.2	Partnering – strategic alliances	156

10	**Your checklists for success**		**159**
	10.1	Future planning	160
	10.2	Mentoring	160
		10.2.1 Finding your mentor	161
		10.2.2 Mentoring explained	161
	10.3	Build on success	162
	10.4	The story so far ...	163
		10.4.1 Who needs business connections anyway?	163
		10.4.2 Business success is 20 per cent strategy, 80 per cent people	164
		10.4.3 How to distinguish between networking, connecting and relationship building	164
		10.4.4 Remember that R is for Relationships – and more	164
		10.4.5 Why contacts are so useful	165
		10.4.6 Perseverance (and more) will get results	166
		10.4.7 Say it with feeling	166
		10.4.8 Touching emotions and intellect	166
		10.4.9 Recommendations work wonders	167
		10.4.10 Don't waste their time	167
		10.4.11 Check progress	167
		10.4.12 Communication skills awareness checklist	168
		10.4.13 Checklist for developing powerful relationships	168
		10.4.14 Every relationship is different	168
	10.5	And finally ...	169

Index   **171**

Preface
The nature of networking

You are probably familiar with the expression, 'People do business with people.' Many professionals know how important it is to be able to deal appropriately with others where business and career development are concerned. Despite the amazing progress of technology, which enables machines to do almost anything you want, instantly, there is still something reassuring and supportive about personal contact and individual attention.

People who work in the fields of science, engineering and technology often have little time to build powerful and valuable business relationships. Their expertise lies in hard skills. Sometimes, due to pressure of work or lack of opportunity, the soft skills are sidelined. When people have little practice in building rapport with others, it makes it even more difficult to find the time to do it. But networking or relationship building should not be underestimated. It is only when people skills are equally balanced with technical ability that professionals can progress their careers swiftly and successfully. Corporate connections are what everyone with ambition should actively seek and nurture. Whether yours is a large or small organization, it is well worth the effort. But how do you go about it? Networking, creating business connections, building rapport in the workplace, developing professional relationships – call it what you will, this book will help you.

So why do workplace relationships matter? What do people who network actually do? For one thing, it does not depend on how clever you are. It is not a matter of intelligence or qualifications. Neither has it anything to do with background, where you were educated, or what part of the world you come from. It also has nothing to do with being male or female. Both men and women can be equally good (or bad) at networking. Finally, where rapport building skills are concerned there is no upper age limit. Anyone can do it if they are minded to, though not everyone does, of course. It is a fact that some people are better at doing it than others, and maybe they are the lucky ones. But everyone can acquire the skill of networking and building rapport with others. All it requires is a bit of practice and a degree of curiosity. Are you ready to read on?

For those reading this book who are already experienced networkers and have many good contacts, it should inspire you to try new approaches or refresh your methods. If you are new to your job or starting out on your career, it is designed to help you make the most of the opportunities offered as you meet people in the course of your work.

1 How to progress your career and acquire valuable contacts and strong allies along the way

Many people know that the power of personal connections helps businesses succeed. In fact, statistics suggest that a staggering 97 per cent of professionals believe it's *whom* you know, rather than *what* you know, that's important. Business today relies as much on people skills as on qualifications and experience. No matter how brilliant you are at your job, if you want to get ahead, good relationships at work will help. Task awareness is fine, and being good at your job is eminently desirable. But, if you can harness that with being passionate about people, it will take you further, faster. Those who tackle this area and get it right have the opportunity to positively influence the growth and profitability of their organization and enhance their career. Achieving your goals is far easier with (to paraphrase the Beatles' song) 'a bit of help from your friends'.

Do you believe that people mean business? I hope you've answered yes. There are two main reasons why people develop the power of personal connections: to help their organization become more successful and stand out ahead of its competitors; and to help themselves progress further and faster along their career path. Because there are few jobs for life, workers frequently switch roles and careers these days. People develop portfolio careers and have many transferable skills. Those who are able to use personal recommendations when moving to new positions often find it easier than people who have to rely solely on their CV. But can you really find opportunities to network anywhere and everywhere? And, if so, what are the benefits?

Developing successful relationships at work means two things: internally (within the organization or profession) and externally (among clients, work providers, suppliers). In both cases, it helps to be confident and have an inherent curiosity. Remember that the same approach is not right for each contact, much less for every occasion. Finding the most appropriate way to further each business relationship is the answer. And this means practising each day, with each person, meeting by meeting. If at first you don't succeed, be flexible and you eventually will. Where relationship building is concerned, it is persistence that gets results.

2 So how do you go about making the right connections for yourself and your company?

Building effective relationships in the workplace is a valuable skill for all professionals. By developing and nurturing a strong network of personal contacts, you will be more effective in business and in your career. With good relationships in the workplace, you are more likely to be successful when taking charge of new situations; also, those with healthy personal networks have higher job satisfaction. Should you have a problem to solve, you may find a solution more swiftly if you have a number of reliable contacts you can turn to for advice. Remember – it's reciprocal. If you help others, they will help you in return.

Initially, you may not feel confident about building business success from professional relationships. This is usually due to having to move into unfamiliar territory. Navigating uncharted water can make even the most confident people feel rather apprehensive. Just remember that developing relationships with different people in the workplace is not difficult once you try it. It is the intelligent approach to business and career development. If you show curiosity about others and they realize you're genuinely interested in them, people usually respond positively.

At each level in an organization, powerful personal connections underpin business success. The formal advantages of good personal connections include colleagues working in cooperative teams; managers whose staff are more productive through encouragement, motivation and guidance; effective directors who provide inspiring leadership; and customers who enjoy doing business with you and are happy to recommend you.

There is no doubt good business connections take time to develop. It's not a quick fix. Tasks can often be accomplished relatively quickly, but, where people are concerned, you can't rush things. Remember, it's people, not skills, education, qualifications or experience. People do business with people they like, trust and respect. Powerful business connections don't necessarily come from people in high places. Sometimes a small piece of information can make a huge amount of difference and you often get help from the most unlikely sources. If you add the power of personal connections to your own individual skills, you will enjoy better, innovative and more profitable relationships with existing and new customers, suppliers, staff, colleagues and superiors. After all, harmonious relationships at work – both inside and outside the company – are what everyone desires.

If you seek to get on better with others, with a particular emphasis on workplace relationships, this is the book for you. It is not uncommon to encounter difficulties in dealing with people who are in a position to influence your career progression. There is plenty of advice on acquiring relationship-building skills, which any professional will find helpful. It also addresses issues that are commonly found in the work environment in a user-friendly and down-to-earth way that is simple to understand. There is little jargon, and management-speak is avoided.

This book also contains guidance on how to deal with the wide range of personalities you meet at work. How to cope with difficult characters and become an ally rather than a foe is also covered. Additionally, there are a number of approaches to help you to rise to the top by offering smooth and subtle strategies to surmount any problems you encounter along the way.

Whatever your level of seniority, you will find tips and tactics to avoid personality clashes and build relationships with people who work for you. At the start there's something about you, the reader, as an individual. This is important because if you don't have a successful relationship with yourself, how can you possibly have a healthy relationship with anyone else? The rest of the book relates to 'other people', who they are, what they're like, how to develop rapport with them. There's an important bit about building relationships with insiders, those within you organisation, colleagues, co-workers, bosses and staff because you need allies in the workplace. And of course there's a lot about relationships outside the company – clients, customers, consumers, influencers, referrers, gatekeepers and much more.

This book is designed to make it possible for all your professional relationships to flourish. It should help your career progress and enable you to develop congruent communication skills. Harmony is what is needed to give everyone an easier life – particularly at work. Rapport building is the core element. Whether you are experienced or just starting out, you will find common-sense tips, advice and suggestions on how to build your own personal network of connections. Armed with this knowledge you will have greater confidence and progress further and faster with your career. If building valuable business relationships is something you would like to do, please continue reading.

Chapter 1
What is meant by business relationships?

1.1 Introduction

Whoever you are, whatever you do, if you can build healthy professional relationships, you will stand out as an impressive member of your organization. You may be newly qualified and in your first job, have recently started a new job in a more senior position or be well on your way up the ladder in your career progression. Whatever your situation, you will find things work more easily if you can get along with other people. To do this, it helps to know a little about what type of person *you* are. If you hold a supervisory or management position you will need to be in control, not only of yourself, but also able to manage your staff and deal with a number of other stakeholders. Being an effective networker isn't something that happens by accident. This chapter deals with some of the points you should consider about yourself, before moving on to building relationships with others in the workplace.

1.2 A bit of self-management

Climbing the slippery pole in any profession is not easy. There is much hard work ahead of you and you will spend many hours in the workplace. Wouldn't it be nice if the majority of that time was enjoyable? While this cannot be guaranteed, you can take some steps towards making sure that your job is closer to a dream than a nightmare. Here are a few points to consider.

If you are starting a new job, what sort of probationary period do you have to serve? The purpose of this is usually to give both you and your new employer time to decide whether you are right for each other. You may be well qualified and more than able to do the job, but, now you've started, do you really like it? Is it what you thought it would be or what you want? There could be a number of factors. You may, for instance, not have enough (or any) autonomy. Maybe size matters. Do you like working for a large organization? Perhaps you do, in which case you'll need to become adept at surviving office politics. Should you hate the idea of the corporate culture, maybe you would be better suited working for a smaller company, where your talents will quickly be noticed.

So, from the outset, pay attention to how you feel in the workplace and how other people react towards you. If you're the least bit unsure, use your probationary time wisely. Give plenty of thought in those first couple of months (sometimes known as the honeymoon period) to finding out whether you will be happy and your line manager will be happy with you. It's amazing how quickly that time passes and any initial dilemmas are best sorted speedily. No problem ever got better by being ignored. Give serious consideration to how things are working out, to make sure you are in the right place to start developing your business relationships.

Before you are able to shine at work and make the best use of your opportunities, consider the likely expectations of the people you'll be working with. Everyone prefers to work with people they can get on with. An ideal colleague would be someone who:

- is positive and enthusiastic;
- is able to see the big picture;
- is capable of achieving their own goals;
- is well organized and self-disciplined;
- is a good decision maker;
- provides honest feedback;
- is fair and doesn't have favourites;
- is open-minded and curious;
- is a good listener (and is available to listen);
- knows and takes an interest in colleagues;
- is a good communicator;
- shows confidence and gives credit;
- keeps people informed;
- acknowledges their own mistakes and weaknesses;
- shares experience and helps others.

> **Networking hint**
> Make it your business to discover what is most important to your colleagues. It will be time well spent.

How many of these attributes do you possess? Be honest when you think about it. Let's hope the people with whom you work closely have some of them too. These skills are the ones that people who want to network successfully should aim to develop. It is an important list, and these skills will crop up from time to time throughout the book.

1.3 Making the right impression

If you're a new employee, or have recently acquired a new job because of your skills, experience or ability, try not to give out too much information about yourself in the beginning. You will learn far more by saying less. Best advice is to keep quiet and listen. Smile and let others do the talking (to you). You might try to figure out the office hierarchy by observation. It certainly won't be the same as is shown on the organizational chart. Read anything and everything you can lay your hands on in the early days, should time permit. When joining a team, treat everyone equally unless there is a clear leader of the group. If you are given the opportunity to have extra training, accept it. Everything you learn is going to come in useful one way or another.

You may be starting a new job, but what if it's not your first? Are you being promoted by your current employer or moving on to a new one? Assuming it is the former case, you should remember that people already know you. But your position in relation to others will, and should, change. You are actually moving on while staying put – if that doesn't sound like a paradox. Because of your new role, you may have to create a bit of 'distance' between you and your former colleagues. Existing relationships and friendships shouldn't dictate the way things will work in future. You may be part of the same team, division or department, in which case you'll need to give consideration to how you act in future. Don't automatically abandon old alliances because of new circumstances – they will still be useful.

Should you be starting work with a new employer, and a new organization, your learning curve will be much steeper. Everything will be unfamiliar. Discretion and caution will be the best tactics initially. Try to match your approach to the actual circumstances. Be realistic about the situation you are in.

Once you start your job there will be lots to get your head around. One good piece of advice is to try to get a meeting with your manager early on. Even if it is just to confirm your role and the priorities. It is well worth setting up an effective communication procedure between you as early as possible. You will be able to find out whether he wants you to report to him regularly. If so, what method is best? Be sure to ask appropriate questions – ideally ones that demonstrate your knowledge and intellect. The purpose of an initial meeting is to help you both make the first few days go smoothly.

If you are able to arrange to be introduced to other key people, it will be helpful. Your work is bound to involve contact with others. They could be in another department; they may be above or on the same level as you. If you can do a bit of research and come across as being fairly well informed about them, what they do and how they fit into the organization, this will show that you have a positive approach. It will help you to cultivate a good working relationship from the outset. When you're new to a position, don't be afraid to ask questions. It is probably the one and only time when most people will be prepared to offer you advice and information. Find out what issues are important to them and what they would most like to get out of working with you.

With a bit of preparation, you will create impact wherever you go and with everyone you meet. This will get you off to a good start. Put simply, this means:

I	Include yourself with colleagues in social activities outside working hours
M	Manage your time effectively while at the office
P	Present yourself well (appearance, body language, voice)
A	Ask appropriate and intelligent questions, then listen to the answer
C	Contribute ideas if invited to do so
T	Think – pause for breath before your speak

You will need to allow time to get to know your colleagues – how they work, their strengths and weaknesses. This can't be done in five minutes.

> **Networking hint**
> While some tasks can be accomplished quickly, people-related activities take a little longer.

When properly managed, working relationships can be extremely rewarding. Don't be put off if some people are particularly nervous of change. They may be shy, be insecure, feel threatened by newcomers or be envious of your success. You may be feeling exactly the same. Should you be working with someone who has held your position (or a similar one) for a long time, they will want to show a bit of control, no matter how senior (and new) you are. Observation and information-gathering are crucial. You must watch how other people work and interact. If it seems that one or two colleagues are being difficult, they are probably just trying to make themselves noticed. It's quite possible that, once they are more used to you, they'll feel less threatened and calm down.

The reason it is essential for you to establish good working relationships is that, so often, you will be reliant on others to help deliver projects, or achieve results. The first step is not to annoy other people just by the way you come across. Good working relationships take time, because you are trying to develop trust, earn respect and build up confidence. If possible, show colleagues from the outset that you are 'on their side' and keen to understand where they are coming from, they are then more likely to become allies and possibly in time friends.

> **Networking hint**
> Avoid making and acting on unwarranted assumptions about people early on. Suspend reaction in order to avoid being judgemental.

Another important thing to remember, if you want to make a good start at building relationships with others, is to be (personally) well organized. This should give you a head start in everything you do. Productivity, effectiveness, hitting targets – all important aspects of your work – are improved by good strategy, preparation and planning. Anything less hinders achievement, and promotes a view that you are not efficient, led by events rather than directing your own destiny. Here's a **smart** approach:

- **S** Set task times: Divide your day/week into sections. If on Mondays you want to be at your desk, avoid meetings that take you out of your office. If you like Fridays to catch up with end-of-week tasks, block out the time to do this.
- **M** Make goals: Clearly defined objectives help focus the mind and keep you motivated. Avoid setting yourself unachievable deadlines.
- **A** Ask for help: Never muddle through. Delegate anything you can. Enlist expertise of others whose skills complement your own.
- **R** Reflect, rather than react: Avoid committing to anything until you have all the facts – a hasty decision could lead to unnecessary stress.

T Think – use your brain: Never be afraid to leave a task if you are stumped. As with exam techniques, if you don't dwell on it but switch to another task, by the time you return to the problem your subconscious may well have a solution.

> **Networking hint**
> If you can – plan the work and work the plan. You not only need a plan, you need to develop a method for personal organization and smart networking.

1.4 Prioritizing

The ability to prioritize is essential. It is what all successful professionals should be able to do. Effective people develop the habit of doing things they *don't like*. If you can work out the important from the unimportant in your new position, you will feel more in control and work more efficiently. Important things require quality time. Urgent things have to be done quickly, otherwise problems will result.

> **Networking hint**
> One senior executive's method for deciding which are the most important tasks for him to tackle is simple but effective. Each day he makes a list of things he wants to get done. He divides it into categories 'A' and 'B'. He tears the list in half. He puts the 'B' list into the waste bin and keeps the 'A' list. He then divides the 'A' list into A and B categories and repeats the process. After three attempts he finds the matters most urgently requiring his attention and deals with them. This may seem a bit drastic for some (though by all means give it a try), but it works.

If you are used to keeping lists, and work from a to-do list, assign each of your tasks into a category such as 'urgent', 'important', 'non-urgent but important', and, finally, 'neither urgent nor important'.

- If something is both urgent and important, put it to the top of your list.
- Deal with the urgent jobs fast, because they are the firefighting, crisis-management things.
- Spend as much time as possible on 'non-urgent but important' tasks, since they are the ones that have the most impact on your work or business.
- Most of the things in the 'neither urgent nor important' category are best outsourced or ignored.

To be well-organized and efficient you should strive to make the most of your personal strengths. Make sure you behave in an appropriate and professional manner at all times – be polite and punctual and keep up to date with your work. Every aspect of your work can be made more effective if you organize yourself and prepare efficiently. It may need some thought; it may even be difficult. Being self-disciplined is a good habit to develop early on. It will help you as you go on to develop professional working relationships.

Successful integration into the workplace where you are new to a job, or to the profession as a whole, requires thought and preparation. Find out as early as possible what the dos and don'ts are in your work environment. If there's a staff handbook available (electronically or in hard copy) read it and become familiar with it. If in doubt about any aspect of it, ask the HR department for clarification. Start as you mean to go on, with a positive attitude and the intention of doing the best you can. Once you are confident about who you are within the organization, you can look towards building relationships with others.

1.5 Self-management and being effective

No serious professional ever strives to be ineffective. But why are some people considered effective and command respect and loyalty, while others are not? Being effective isn't just about being organized, or self-disciplined. Nor is it about getting your team to deliver projects on time and within budget – even if those are elements of your work on which you can be judged. An effective person is highly motivated and keeps the broader picture in mind while inspiring colleagues and employees to excel in their work.

For some this comes easily. They have natural charisma and style people admire. But for others, however competent they are in their professional expertise, they do not easily command respect, loyalty or trust. They need to develop their own natural communication style before they can begin to develop workplace relationships. Being effective is based on the ability to keep an open mind and learn from others. If you are ready to accept responsibility and be accountable for your own actions, you are likely to be able to build effective networks with others.

If you are at the beginning of your career, working long hours, trying to deliver a difficult project or reach performance targets, you are probably doing the best you can. But do you know how you are viewed by your colleagues, staff, partners or directors?

1.6 Getting the best out of feedback

Why not try a bit of informal feedback through the team and departments you are working with? You may be encouraged by what you hear. If the results are a bit negative, don't worry. Have a look at the words in the left-hand column below. Are any of these (or similar) likely to describe you? If so, you may need to work towards being more like the words in the right-hand column. That encompasses positive attributes. The more ticks you get in the right-hand column, the better. For future reference, when trying to establish rapport with other people, it is harder if negative behaviour is their norm.

Negative – ineffective	BEHAVIOUR	Positive – effective
Evasive	E	Encouraging
Faltering	F	Forward-thinking
Frustrated	F	Fun
Enigmatic	E	Experienced
Critical	C	Confident
Troublesome	T	Trustworthy
Insincere	I	Interested
Visionless	V	Visionary
Egotistic	E	Enthusiastic

You may be wondering why this issue is being described here. Because it is so important, that's the reason. Should you not be aware of the negative impact of your or someone else's behaviour, the sooner you pick up on it the better. If the behaviour isn't yours, but someone else's with whom you work, being able to recognize it will help you to avoid picking up those traits yourself. Perhaps you can recall colleagues with whom you've worked recently who you'd regard as ineffective. Do the words in the left-hand column here seem appropriate when describing their actions?

Negative – uninspiring	BEHAVIOUR	Positive – inspiring
Indifferent	I	Imaginative
Nervous	N	Natural
Suffering	S	Sympathetic
Pressured	P	Punctual
Insipid	I	Impressive
Rigid	R	Relaxed
Embarrassed	E	Efficient

Consider the words connected to EFFECTIVE and INSPIRE above. If you were to assess your behaviour at work – negative and positive (left-hand and right-hand columns) – how do you think you would rate? What do think those around you, who work with and for you, think about your attitudes and approach in the workplace? When seeking feedback, you will need to consider what insights it has given you, what surprises you got (if any). Were they good, bad, ugly or interesting? Which areas do you need to improve?

> **Networking hint**
> After receiving feedback, you should have a better idea of your talents and skills, areas where you are most appreciated, areas you need to freshen up and skills you should consider learning.

Remember what was said in the beginning: the ability to network and build relationships with other people doesn't depend on whether you are a first-jobber, experienced professional or the head of a department or company. Everyone will benefit if they possess skills that will help them do their work more effectively and work well with other people.

Going back to self-analysis, some questions you could ask yourself are:

- How do you like to be managed?
- What skills do you need to be able to do your job?
- What support would help you?

To be effective, you (and your team if you are part of one) need to be focused and working with purpose. If you are currently involved on a project, how would you classify yourself and your team?

- High-focus, low-energy = disengaged.
- Low-focus, low-energy = procrastinating.
- High-energy, low-focus = distracted.
- High-energy, high-focus = purposeful – positive results.

Some people are driven to perform 100 per cent in the workplace, and expect everyone around them to sustain that level of effectiveness as well. Over and above that, sometimes you will have to manage yourself and inspire others to give 100 per cent of themselves. This is a tough one to deliver, unless you're able to show by example that you always give the full 100 per cent of yourself to your job. This isn't a question of staying at the office for extra-long hours, working over the weekends or being too busy to take holidays.

Networking hint
To sustain the high-energy, high-focus approach you will need to arrive at the best means of ensuring you work effectively during working hours while leading a balanced life outside the workplace. This will gain you the respect of your colleagues and put you in a good position for building positive professional relationships.

1.7 Seeing the bigger picture

Any senior professional needs to know how to amalgamate the successful running of a project or department while being alive to opportunities for growth or new business. In other words, you should harness the experience and knowledge of the past to help deliver the present and future objectives. It is important that you hold in mind the broader picture of cross-company objectives and future business aspirations.

Effective networkers acquire the skill of collaborating internally and externally. You will learn to build relationships and forge alliances of interest with others – both inside and outside the organization. Once you start doing this, you will work actively to promote good cooperative relationships – whether at an informal meeting or on formal occasions.

A useful skill that should be learned as early as possible is how to make effective decisions. You don't have to be a head of department to need to do this, although, depending on how high up the organizational chart you are, it can sometimes mean the decision-making process is a lonely one. Everyone does it in their own way. If you think you're not all that good at making decisions, practise on low-impact issues. You will become more accustomed to decision making the more you do it. How do you currently make decisions? Are you likely to use one style more that another? Do you need to take a more balanced way of decision-making in the workplace? Research has shown that there are a number of styles used to make decisions.

- *Logical*

You weigh up the evidence and make a decision based on the facts alone. This is a traditional analytical business process of making decisions: 'There are X, Y and Z options available to us at this time.' This is a normal decision-making style in a crisis situation or an emergency. It is, however, a somewhat limited basis for a decision.

- *Intuitive*

This is where you make a decision on gut instinct. It is a well-known fact that lasting impressions are made in the first few seconds of meetings between strangers. Whether this is true or false, if you use this method to make your decisions, those instinctive first impressions will have a lasting influence, which may in time prove unfounded. Very often used where relationship building is concerned.

- *Compliant*

Do you make decisions that just 'go with the flow'? These decisions are the ones that are taken just to maintain the status quo or to keep others happy. Try not to confine yourself to this method: it could be construed as a no-thought decision since there would have been little debate and it might be the weak option.

- *Hesitant*

These decisions are the ones that just don't get taken. You will procrastinate and put off making up your mind till the last possible moment. The most experienced practitioners become so adept at this skill that often someone else makes the decision for them in the end. Sometimes circumstances change, so no decision needs to be made after all. If you hesitate because you simply don't know what to do, take a decision, then make it work.

- *No-thought*

This comes about because whatever the decision, the outcome is of no real interest or benefit to you or your colleagues. You simply leave the decision-making process to other people on the basis that you will rubber-stamp whatever course of action they come up with. This may be an empowering approach on the one hand. On the other, it can also cover up an indecisive manner, adding pressure on you in future, should someone in higher authority question why that particular action was taken.

If you want to be an effective networker, you will need to be able to take decisions and make them work. Practise this skill whenever opportunities arise. It will make your confidence grow when dealing with other people.

1.8 How do you come across?

If you want to deal effectively with other people, it's helpful if you can avoid complications in communicating with them. Misunderstandings can and will occur, and in most cases it is because of a lack of clear communication. Most people assume it is the other person's fault. If you are seeking to build positive rapport with other people you should communicate in the most unambiguous way. It is possible that it is the other party's fault, but could what you say, your manner or style of speech, cause a misunderstanding?

Communication is the successful transmission of an idea from one person's mind to another's. It is a process fraught with obstacles. All sorts of problems can occur here, such as:

- a lack of concentration;
- a perceived prejudice about the communicator;
- false assumptions about the message;
- dislike of the communicator.

To be clear and unable to be misunderstood is the ideal. If someone wants to misinterpret what you are saying, there are plenty of opportunities for them to do so. It is not surprising that communication often goes wrong. One study showed that, on average, people leaving an hour-long business meeting had three to four major misconceptions about what had been agreed.

In order to build good relationships with others, you will want to avoid such occurrences. Have you ever been in a situation when dealing with other professionals where you've hit a communication barrier? This can be extremely dangerous. You may have invented reasons and excuses to convince yourself that the problem had nothing to do with you. Success or failure where communication is concerned is often the result of the attitude of the individual. Whether you are the 'transmitter' or 'receiver', a change of attitude can bring about an outstanding reversal of results.

The most successful communicators are deeply motivated individuals. It is rare to find successful individuals who have become successful by doing what they hate or dislike in life.

Where there are problems of communication, it is often the recipient of the communication perceiving it to be negative that causes the upset. You may consider someone a difficult person if they are delivering ideas that you either don't like or don't agree with.

In a situation where the automatic reaction is to feel negative towards what is being communicated, remember to:

- pay attention, listen, duplicate and understand;
- make no assumptions;
- listen for any free information;
- acknowledge their ideas, repeating the essence if necessary;
- ask a question to ensure the communication is clear or satisfactory; if it isn't, identify the expectations before concluding the exchange.

If you communicate with people fairly, in an open-minded way, there should be less opportunity for misunderstanding. This may not always be the case, but you should aim to be responsible in the way you communicate with others.

Clear communication comprises:

- showing interest in the other person by using their name;
- showing empathy: remember your own experiences of when you have been misunderstood or needed clarification;
- considering the possibility of human error: you don't want to be misheard or misrepresented, so make sure they have all the facts;
- taking responsibility for the communication.

If you are trying to explain something which is difficult, keep language as simple as possible.

1.9 Managing other people's objections

Remember, when communicating, people won't always agree with you. Take each and every opportunity to listen and acknowledge. Effective and positive communication begins with recognition and appreciation that each party has a right to hold different views. It is helpful if you can tone your body language and voice to be as neutral as possible.

All of the above is achievable, but the first step towards mastery of rapport building and effective communication skills is to know what you want out of any communication. Once you have this established you will need three skills to work with the process.

1. *Sensory acuity and awareness:* Become more alert and aware to the responses and actions of others. See more, hear more and feel more. This is a skill that can be learned.
2. *Flexibility:* If you are not getting the response you want, you need to be able to change your behaviour, body language, tone of voice, phraseology, until you get your desired outcome. You cannot demand that the other person or people change.
3. *Congruence or authenticity:* This simply means that what you say and how you say it convey the same message.

> **Networking hint**
> An atmosphere of trust, confidence and participation is essential when building rapport or empathy with other people. They need to feel able to respond freely.

Communication flows when two people are in rapport, when their bodies as well as their words match each other. What you say can create or destroy rapport but that accounts for only 7 per cent of the communication. Body language and tonality are more important. People who are getting on well tend to mirror and match each other in posture, gesture and eye contact. Their body language is complementary.

Successful people create rapport and in turn this helps to develop trust. This is achieved by consciously refining your natural skills. Through matching and mirroring body language and tonality, you can very quickly achieve a bond with someone. This matching must be done sensitively and with respect at all times.

Your beliefs and interests also condition what you notice. There are three main representational systems or modes through which people access, store and filter their experience. These are *visual*, *auditory* and *kinaesthetic*. You operate in all three modes at different times, but tend to have one preferred mode.

First, you should try to develop the skill of being aware of others' moods. Then be flexible in your own method of communicating. People who are visually oriented will pay attention to what they see when you are speaking; auditory people will hear you and listen acutely to the words and language you use; the people who are kinaesthetically inclined (i.e. sensitive to bodily movements and how they actually feel – sometimes called 'muscle sense') will try to tune into your thoughts and feelings via non-verbal cues in order to understand your ideas.

1.10 Awareness of others and flexibility

In order to control communication with others, you will need to be able to:

- identify the style of communication that is preferable to the other person;
- understand accurately what they want – not what you think they should want;
- effectively manage and respond to potentially emotional situations;
- observe the impact your communication is having and alter it accordingly.

Awareness is achieved through the use of the senses and is a skill you should learn. You literally learn to see, hear and feel more.

It is easy to talk with someone with whom you feel you're 'on the same wavelength'. Flexibility enables you to alter your mode of communication to suit whatever situation you are in. In order to 'speak the same language' you may have to change your usual form of delivery. By using flexible communication skills you can build rapport with people, and it is a useful skill when having to deal with challenging individuals and situations. People like people who are like themselves.

This is the end of Chapter 1, which has considered you – you as an individual. It has covered aspects of self-analysis, personal organization, self-discipline and communication skills.

Chapter 2 covers how to begin building relationships with other people.

Chapter 2
Why do people network?

One of the points made in the last chapter was: People do business with people.

The purpose of building rapport with others is that you will be remembered favourably the next time you meet. Developing business relationships requires time, commitment and effort. You need to be focused and self-disciplined, and have patience. When people like you, they will probably be happy to do business with you – it's that simple. Rapport building is about empathy and persuasion.

2.1 R for relationships

Why is it beneficial to build relationships with other people? The two main reasons are: it's practical as well as profitable; and often the most successful people are the best connected. There are some other good reasons, which are listed below.

- Recognition: Creating the right impact when you meet people, particularly for the first time, is something that works well. Never underestimate the power of first impressions.
- Recall: If someone can recall you easily to mind, you've made (hopefully) a favourable impact when you first were introduced.
- Reaction: One of the things you are hoping for is a positive reaction when you encounter them again.
- Respect: Aim to gain their trust. The ability to cooperate with and assist others is vital. You will then earn respect. Don't forget to show it to others in return.
- Responsibility: Do take responsibility for your business relationships. That will keep you in control of the personal network you are going to develop. It represents a lot of effort and will become an invaluable resource.

> **Networking hint**
> Everyone has their own group of personal contacts – their unique network. How many people are in yours? Do you value it? How do you use it?

Some ways in which your personal network can be used are: as a **research aid**, for information gathering; as a **link to new contacts** or markets; and to **advance your career** by meeting interesting people.

2.2 Making a start

Building powerful business connections is a three-stage process. First, you **network** to make connections. Then you use the **connections** to build relationships. Finally, your **relationships**, used wisely, bring rewards. (The way to do this will be described in detail as we move on, beginning with networking at 2.3.1 below.) The list of positives that can be achieved by having rapport with others is extensive.

By harnessing the power of personal connections, you will appear to advantage within, as well as outside, your organization. You can chart the success of your own goals and objectives. One of the biggest bonuses is that you will get to meet like-minded people.

Most people have a network of around 250 contacts – and within that there are perhaps about 25 individuals whom they know really well. You can develop the network you most want, and identify those with whom you wish to build rapport. It is infinitely flexible and adaptable. Your objectives may alter as you progress. But you can change the links and paths to achieve your plans. You can focus on success in both business and career development.

Business effectiveness increasingly depends on interpersonal skills, and creating trust is paramount. Once you begin to have successes you will be able to measure results. Put people first – they are the most important ingredient of all.

There are two important things to remember: networking works both inside and outside an organization. The purpose of building good relationships within your company is to be well informed. This saves valuable time and increases productivity. You also pick up on internal politics and are able to maximize opportunities that come your way. Externally, you will have valuable contacts who act as referrers, bridges, sources, links and influencers to help you achieve your personal or corporate goals. The key skills required, if you don't already have them, are enquiring, listening, researching and organizing.

Here are some examples of companies that invested time and effort in building the right connections. Each one describes a different form of networking.

2.2.1 Business introductions

A newly formed IT-based company wanted to find potential investors. They worked hard to develop a network of business connections to help them source potential funders. They spent six months researching suitable prospects, influencers and connectors to identify a number of venture capitalists and investors.

From a number of introductions, they found three venture capitalists who were prepared to help them. They now have sufficient investment for their growth over the next five years. Their business is on track to perform to its maximum potential.

2.2.2 Strategic alliances

An international firm of consulting engineers wanted to improve their ability to win more business in the UK and Europe. Their business-development strategy was drawn up to increase their links with, and standing among, the most relevant movers and shakers in the construction and property industry.

Why do people network? 17

By working hard to raise their profile within their profession, they nurtured influencers and recommenders who helped them build strong strategic alliances. They have recently won a number of important construction awards and international acclaim for several projects.

2.2.3 Client relationships

A high-profile UK architectural practice carried out a client-satisfaction survey to benchmark their reputation. They compiled questionnaires and interviewed a selection of clients, prospects and influencers.

The feedback was positive. The results reflected the company's strengths and identified areas where improvement was necessary. By following up on some of the encouraging comments received, the directors were able to acquire repeat business and extension work, which increased their annual turnover by almost 20 per cent.

2.2.4 Team building

A charity were organizing a high-profile fundraiser at a time when there were a number of other, similar events. Apathy had set in, take-up on tables was slow and ticket sales were sluggish. The organizers needed to harness energy and enthuse staff to take the project forward.

By calling in favours from some of the charity's high-profile patrons and supporters, it was possible to rekindle enthusiasm and energise volunteers. A brainstorming session took place among a group of individuals who had diverse skills and personality types. A number of new initiatives were suggested. Each person was designated a role that most suited their personal strengths. The result: the project went forward with eagerness because volunteers worked harmoniously and productively together and the event was a sell-out.

2.2.5 Succession planning

A global financial institution needed to help some of their executives who were on track for career progression. Although successful within the organization, a number of these individuals had been appraised as being too task-oriented. Due to their reluctance to develop personal contacts within the company and their weak interpersonal skills, their promotion prospects were on hold.

The company looked within itself and, through its internal network, identified several senior managers who acted as mentors for these young managers. By setting up this coaching programme, the company were able to help them back on track. Their interpersonal skills increased, they became less task-aware and the company gained the advantage of smoothing their succession planning.

2.3 The three-stage plan: how it works

Our three-stage plan encourages networking (described in this section), making connections (discussed in Chapter 3) and relationships (covered in Chapter 4).

2.3.1 The importance of networking

Here, we will take a long look at networking. Are you ready to take action? Remember that, if you can harness the power of personal connections, it will enhance your career prospects and help your company's business development. Your goal is to make quality contacts and devise a plan for strategic relationship building. The process starts here. You are probably familiar with the networking process even if you do not agree with it, nor have time for it, nor feel it is of any great purpose.

> **Networking hint**
> A useful mnemonic to help build business connections:
>
> **R A P P O R T:** Relationships
> **A**re
> **P**owerful
> **P**roviding
> **O**pportunities and
> **R**ewards
> **T**oday

Rapport building requires the development of a genuine interest in other people. You could call this a natural curiosity. Those who are able to cultivate such skills have great advantages over those who don't. Remember to nurture any professional relationships you want to develop. This involves getting to know people and introducing them to others, so that there is constant movement forward. There are many businesses, professional associations and other organizations that offer networking opportunities –breakfasts, lunches or evening receptions. It is an opportunity to meet people for business-development purposes. But what you do *after* that initial meeting is what this book is about. One of the main reasons people network is to create new business opportunities. When you want to build a relationship, a good way to start is to ask the right questions. It's far more important to be interested, than interesting.

Networks matter – it's as simple as that. They are part of the corporate survival strategy. Networks enable you to access work, resources and opportunities. They also create a sense of community and rapport and allow you to share experiences with like-minded people.

> **Networking hint**
> Why is networking essential? Whether you are a self-employed professional or ambitious to succeed in your company, industry or sector, the 'need to know' and 'need to be known by' are important business skills.

If you want to assemble or improve your own network, you need to review your existing contacts to see if they are effective and current. How would you answer the following questions?

- Do you communicate regularly with your own network?
- Do you proactively seek to increase and refresh your contacts?
- Is it a formal or informal arrangement?
- Have you asked your contacts for help or offered support and advice to them?
- Do you keep your network in good shape for easy access and management?

Networks function best according to the amount of give and take. You get out only what you are prepared to put in. The best networks are information-rich, collaborative, high-trust environments. To be part of a vibrant network it is best to start simply. Some of the best ways are through:

- personal contacts, friends and associates;
- ex-colleagues and present co-workers;
- alumni networks;
- clients and professional contacts;
- professional associations;
- national umbrella organizations.

These should be a rich source of opportunities. Most successful networks operate on the basis of personal introductions and referrals. For those who want to refresh their networking skills, or have one or two confidence issues, the following key pointers can be used as an *aide-mémoire* before moving on to the next stage.

2.3.1.1 Get yourself organized for the event

Preparation is essential. Proactive business relationships don't happen just by chance. First and foremost, have a plan. This really matters. If you don't know why you're doing something, you won't do it well.

If you're a bit reluctant to **move out of your familiar territory**, preparation helps. The more industry functions and business events you attend, the less threatening they will be. To have a vibrant network you need to keep widening your range of contacts.

When you **attend an event**, do a bit of research beforehand. How important is it? Your time is valuable and you should always make the best use of it you can. Should you be accompanied by a colleague or do you go alone? Is one person sufficient? Would two heads be better than one, if there are a lot of people to meet?

Study the delegate list beforehand, if possible. This way you maximize the opportunity to reach the right people.

Ensure that you are appropriately dressed. Always check the **dress code** – appearances can make or break an encounter.

Correct location? Some business venues are much like any other and you can waste hours looking for them. Are you at the right event? Even some conference suites have similar-sounding names. It is possible to find yourself part of the wrong group of delegates.

Be punctual – a late arrival doesn't convey the best impression. Remember to take your business cards with you and have them easily accessible. It conveys a more professional approach.

If for some reason you arrive early, don't worry. **Get involved**. Seek out the organizers and see if you can offer to help. Even if it is politely declined, it will indicate that you are a considerate person. Should you be asked to assist, it will give you something to do and you won't have time to be nervous.

If you hate the idea of networking, you are not alone in this. Many people don't network because they don't like talking to strangers. If it's any comfort, Woody Allen once said, 'eighty per cent of success is turning up.'

2.3.1.2 How to start off

Be positive and outwardly confident, it will make you stand out above others. Don't worry if you're wishing you were almost anywhere else. Research shows that more than 90 per cent of people feel fear about walking into a room full of strangers.

Avoid arriving with coats, cases, umbrellas and other baggage. Leave belongings at the cloakroom so you can **appear calm, unflustered** and unencumbered. Remember, you are judged in the first 15 seconds of meeting someone. Don't waste your opportunity.

2.3.1.3 Smile – it signals confidence and openness

If you walk into the room alone, **look for other people** standing on their own and make an approach. The chances are that they are feeling awkward too. Striking up a conversation with them will show that you care about others and are not preoccupied with yourself. It will help them feel better about being there and you will have made a new contact already.

Introduce yourself by your first name unless it is a really formal occasion. First names are more informal and you will convey the impression of being approachable. When introduced to someone, **use a firm handshake**. Have the name of your host/organization at the ready. It helps if you know whose guest you are, or which group you are expected to join.

2.3.1.4 Making an entrance

To make an impact – work on your opening line. This is often called a 'lift pitch'. Rehearse a sentence that summarises (in about 40 seconds) what anyone needs to know about what you do.

Posture is important – stand up straight. At business functions you should always be alert and attentive.

Look around the room for acquaintances or friendly faces. This doesn't mean latching on to a colleague and cowering in a corner with them for the rest of the evening, avoiding everyone else.

Maintain appropriate eye contact. Avoid staring, glaring, winking, blinking or looking straight over people's shoulders to others beyond. Keep calm and relaxed. If you can take some deep breaths it helps you to be alert and able to concentrate.

Develop good rapport by asking a non-threatening question. Open questions are best rather than something that is likely to produce a yes/no answer. Even an enquiry as simple as 'Have you travelled far to get here?' opens the conversational gambit.

Be curious – when talking to someone, varying your vocal tone will make your conversation more interesting. You can charm people by being sincerely interested in them and listening to what they say.

2.3.1.5 Some things to watch out for

If you can, **avoid contact with potential troublemakers**. For example, there may be a person present who's had too much to drink and wants you to give them a lift home.

It's best **not to lecture people** or use emotive gestures, however passionately you may feel about the subject under discussion. Don't be aggressive or speak with force. You are not at a political rally.

Decisive people don't dither. If you find yourself in awkward company and you want to make a discreet exit from the group, get ready. Become the master of the graceful withdrawal. Make your excuse politely but firmly. Say, 'I'm sorry, but I see my colleague is about to leave and I must speak to him.'

Eating and talking. By all means eat the delicious food that's being handed round, but try not to talk at the same time. It's difficult to engage in meaningful conversation with someone whose mouth is full. Equally, if you are eating when you're asked a question, wait until you've finished before replying.

Rapport is a two-way thing. Dominating people often monopolize conversations with loud, uninteresting details about themselves. Withdraw with tact and dignity. In the same way, if you've tried several opening gambits and have been met by a blank stare and monosyllabic answers, perhaps you should move on.

Beware the predatory person whose sole purpose is to meet and attempt to date attractive fellow guests. (This warning is non-gender-specific.) Also **watch out for limpets**, the nervous types who, having earmarked you as friendly, stick to you throughout the evening and are impossible to shake off.

2.3.1.6 How to be memorable

By **being polite and courteous**, you will be unforgettable. If you can cultivate charisma you will go far. Deal with people kindly and sympathetically. Offer to fetch someone a drink, or introduce a stranger into your group.

Take pride in what you do and **be professional**. Even if you are a bit sceptical or reluctant about the value of networking, whatever the occasion, you never know whom you might meet. Keeping a conversational tone to your voice encourages people to respond to you in a friendly manner.

Speak slowly and clearly. So often in venues, there is loud background noise because of the lack of soft furnishings. Conversations are hard to maintain above this level of sound.

You need not feel insignificant when talking to others who are more brilliant or experienced than you. Remember, emotional intelligence stands out way beyond paper qualifications. **Develop the swan technique** – keep a calm exterior and smooth behaviour. No one can see what's going on beneath the waterline.

Try to establish connections with people – this comes naturally with practice. You could ask a question about the event: Have you heard the speaker before? Do you come from a particular region/country? If there are some people on their own, seek them out by asking if they have seen the host/hostess. You may be equally shy but it will help you overcome your nerves.

2.3.1.7 Some common pitfalls to avoid

Avoid clumsy exit lines. If you're really stuck, try 'I've taken up enough of your time. I really mustn't monopolise you any longer.' Watch your body language and gestures. Don't rush – tension is easily communicated.

If you have a time limit, be polite and excuse yourself with tact. If you have a tendency towards being a 'butterfly', slow down. You can't speak to everyone in the room, so there's no point in trying.

Make appropriate eye contact so as not to unnerve your companion. The eye dart – looking at a person for less than two seconds before your gaze flicks elsewhere – is most unsettling. The opposite – the fixed unblinking stare – is just as bad. Also avoid 'mowing the lawn' – looking from one side of the room to the other. It is disconcerting for those in your group, and impolite.

Check those annoying habits. However tense you may be, playing with the hair, fiddling with jewellery or tie is not acceptable. Any hand wringing, twitching or fidgeting is vastly irritating to watch.

For those who are particularly nervous it may sound simple but **don't forget to breathe**. This simple relaxation technique is often overlooked. Most experienced actors and public speakers practise deep breathing before going out on stage. Several deep breaths will help.

If you are bored, don't sit down. It's too easy to get trapped that way.

Use light-hearted anecdotes as conversation fillers, but always be sensitive to other people's feelings. Initiate a conversation by asking a question. This is a useful rule but make sure the question is innocuous. There is a lot of difference between being curious and being prurient.

Mind gone blank? Ask for someone's business card on the pretext of giving it to another guest by way of introduction. Repeat their name a couple of times (hopefully) to fix it in your mind.

2.3.1.8 Some conversational tips

Try teaming up with a colleague. Do you use the buddy system? It helps to get round the room more effectively, particularly if it is a large gathering. Ask your host to help if there is someone you specifically want to meet. They can arrange to introduce you.

Watch others in action and emulate those you admire. You can always learn from any situation – however experienced you are. If someone else's actions make you cringe with embarrassment, at least you've seen how not to do something.

Keeping your eyes and ears open to what is going on around you heightens awareness. When moving from group to group around the room, do so with purpose but in a controlled manner.

When you chat, **allow your voice to have expression**. Warmth and confidence encourage responses in others. However fascinating you may find one person, don't ignore the rest of your group. Encourage others to talk about themselves or their interests.

Use pauses for effect. Never underestimate the power of silence. You will have your companions' undivided attention – for a few seconds. Don't interrupt when someone else is speaking. This is rude and alienates people.

2.3.1.9 Networking should be fun

Enjoy yourself, the venue and the hospitality. After all, this is supposed to be a social occasion, even though it is work-related.

Mix with the other guests. Don't get glued to one spot or stick to one group. If there is someone you particularly want to meet, make contact at some point during the event. Even if they are about to leave, hand them your card and ask for their email address.

Enthusiasm is catching. Being positive is attractive. It helps shyness evaporate and will get you noticed. It is important to stay in control. Are you likely to get carried away? If so, watch the alcohol consumption.

2.3.1.10 Ways to handle awkward situations

If you make an embarrassing social gaffe, have the courage to admit the fault and apologise. Being upfront and honest can turn a mistake to your advantage.

Avoid a difficult question by asking someone else's view. You could even turn the question round and ask the person posing the question for their response. When you have listened to their reply it may help your own reaction.

If someone drops a bombshell they may be testing your response. Rather than give them the satisfaction of erupting, try to suspend reaction. A measured acknowledgement buys you time and can save faces and reputations.

Making a tactical and diplomatic withdrawal from a potentially explosive situation is sometimes the only way to avoid disaster. Have an escape route handy: 'I've just seen my colleague beckoning me – please excuse me, I must go.'

If you are forced to listen to gossip, steer clear of getting involved in the discussion. Do not get drawn into conversation. Offer advice only when specifically asked for it. No one likes a know-it-all.

Respect others' personal space – stand about a metre away from people you don't know well. Body language will indicate if and when it's safe to approach more closely.

2.3.1.11 Reaping the rewards

The more you attend networking events, the easier it becomes. Your social skills become second nature to you. Continually practise wherever you are – in the supermarket, at the gym, when travelling.

Join other organizations where you can help to increase your confidence. New members are often welcomed in groups – take advantage of every opportunity.

Devise ways of being of value to others. Offer to share your contacts and skills with them. Ask them to reciprocate. Pay attention to each new personal connection you make. Show interest in them, sincerely, by following up shortly after your meeting.

Take contact details when you meet someone you think could be interesting to get to know better. Make sure you have an effective way of keeping notes. This is your unique contacts database. Your memory – however good it is – will fail you sometimes.

Make regular telephone-call follow-ups. Very few people do and it fixes you in people's minds. Call people even when you don't want anything from them. They will be amazed – and they will remember you.

2.3.1.12 Understanding the process

Now that the networking process has been described in detail, perhaps you can see why it is essential. It is important to build internal and external contacts. Networking builds rapport with superiors, subordinates and peers. It also fosters relationships with customers, suppliers and competitors.

Networking and selling, however, are like chalk and cheese. Networking events should be used as a platform to make positive business connections. You may sell yourself, but not your products or services. There is much to be gained professionally and personally from networking inside and outside your company.

At its best, networking is a process of making connections with a diverse range of people. These connections can then be developed into reciprocal relationships to increase your business or advance your career.

One of the reasons why networking gets a bad name is that people who do not understand the process abuse it – by trying to sell services or products. Meeting people who do not respect your values and attitudes, who have poor interpersonal skills and who find it difficult to share, can be offputting.

If you've encountered people who go to an event unprepared, have low self-esteem and fear rejection, you'll know that they require careful handling. Through shyness, their persona and approach are negative. They lack self-confidence and show no signs of curiosity about others because they are too wrapped up in themselves.

Conversely, those who miss the point of networking creatively will attempt to dominate groups and conversations. They will not engage in dialogue or show any interest in offering help to others.

By keeping the positive benefits firmly in mind, you will find the process an enjoyable precursor to the next stages of building rapport with other people.

2.3.1.13 Networking – summary

- Everyone needs to network, whether it is for jobs, information or fresh ideas.
- Have a plan – if you know what your goals are you'll be able to work out whom you need to meet to get there.
- Analyse your network – whom do you already know and how well do you know them.
- Don't assume the most senior people are the most valuable – pay attention to juniors and admin staff, too.
- Attend selected events – you can't be everywhere and you don't need to be.
- Image is important – not too formal nor too scruffy.
- Be organized – when you collect details make a note of them and follow up with information that is relevant to your contacts.
- Be consistent. Do make time for contacting people – if you are persistent the rewards are high.

2.3.2 Developing successful working relationships

If you're starting to get keen on the idea of networking, you may want to spend some of your time talking to colleagues. If your 'need to know' is all-important, there's nothing like gossip to help fill the gaps. After all, until you've heard everything, how can you possibly decide how important the information is and whether you need to know it or not? Being gossip-averse can be short-sighted. Conversation is the way relationships are formed.

Relationships between colleagues are a company's greatest asset. If people can't work together it is far more difficult for work to get done. Nor is it very pleasant having to do it. If you can learn how to integrate well and appropriately, you will be able to hold good-quality conversations with co-workers and be a valuable asset to your organization.

> **Networking hint**
> Conversation should not be confused with communication. Communication is about exchanging information; conversation is a creative process and engages people's minds.

Conversations don't stick to agendas, neither do they incorporate jargon or management theory and hype. Conversation is about connectivity – enabling staff to keep in touch with one another. They are an antidote to stress and other health problems. You should try to cultivate this habit. People who have good social relationships at work are far less likely to be anxious, stressed, absent or seeking to move on.

There is bound to be somebody (or even more than one) at work with whom you will have some difficulty connecting. This could cause a bit of a stumbling block, particularly if this person is in your direct reporting line. There could be a number of reasons for this: disappointment at their lack of progress in the office, compared with yours; unhappy home life; some other reason. One way of tackling the issue is to pause, hold back from making any further overtures to them. If they are continually ignoring or rebuffing your approaches, you have nothing to lose by taking a bit of time over this.

Think about the way you felt about that person when you first met them.

- What initial reaction did you get?
- What sort of voice did they have?
- Did they have a firm handshake?
- Did they readily make eye contact with you?

Anyone new to the networking process should pay particular attention to this. For example, do you listen to the sound of someone's voice? Do you notice their touch as they shake your hand? Is it strong or weak? Are they tactile or reserved? What about their physicality – do they stand near to you, inappropriately close or far away? If you can tune into someone's persona you could reduce the likelihood of making incorrect judgements about people.

2.3.2.1 Building relationships with difficult people

There's no doubt that at some time or other in your new job you will have to deal with some difficult people. One piece of advice I was once given came from a highly successful financial director when I asked him to what he attributed his success. He said he had, early on in his career, developed the combined skills of an acrobat, a diplomat and a doormat. The key to the issue, he believed, was knowing in which order and in what proportion these skills should be used.

Your aim should be to try to work that one out.

2.3.2.2 Assertiveness

When you're trying to build good relationships at work it is particularly difficult if one or two of your colleagues are high-maintenance people (HMPs), or downright bullies. In this situation you should review your assertiveness techniques. The key to being assertive is that, in any difficult situation, you leave it feeling OK about yourself and the other person involved.

The aim is for a win–win outcome in terms of mutual respect and self-respect. The bonus is that there's an absence of anxiety afterwards. You won't have feelings of guilt, embarrassment or frustration.

The difference between being aggressive, passive and assertive is clarified as follows.

- An **aggressive response** is a put-down. It is a personal attack, tinged with sarcasm and arrogance.
- A **passive response** is your choice not to say or do anything confrontational. But it can leave you feeling frustrated afterwards
- An **assertive response** is a reasonable objection that is delivered in a polite and positive manner.

You may well be anxious to please. In this situation, how do you normally respond? If you are trying to avoid being taken advantage of by the HMPs in your office, you will need to be able to think on your feet. If you find yourself in a tricky situation, an assertive response is the best one for you, because there is likely to be a win–win.

Passive behaviour gives you no advantage, and you can lose a good deal from behaving aggressively. But what you can gain from being assertive is that you feel good about yourself. You also have the satisfaction of knowing that you have handled a difficult situation correctly. There will be an absence of anxiety and guilt. Once you have worked out what the tangible benefits are, it will make you more assertive in future.

2.3.2.3 Some situations to consider

SITUATION 1. Your manager asks you to work over the weekend for the second time in a month. You know the importance of the deadline and you are anxious to please. But it's a significant family occasion and you've promised you'll attend the celebrations.

Solution A. You could tell your boss you've done your fair share already, having given up your previous weekend. You mention that it's not fair on your family life and there are other people he should ask.

Solution B. You could resign yourself to working over the weekend. You've got to go home and explain the situation to your family and then spend the whole of the weekend feeling resentful and guilty.

Solution C. You could say you have other commitments but suggest coming in early on Monday and offer to stay late a couple of evenings that week if that would help.

SITUATION 2. You've started working for a company with an established long-hours culture. This is a new experience for you and it's wearing you out. You decide, if you are going to continue with your job, that you need to cut back to a four-day week, so you prepare workable solutions to present to your directors.

Solution A. Your suggestions are turned down so you plead with them, explaining that the way you work is making life impossible.

Solution B. You threaten to resign if they won't compromise.

Solution C. You ask for a detailed explanation from them as to why they have rejected your proposal. Once you've seen it, you realize that they have had a bad experience in the past with someone else who suggested a similar scenario. You rework your proposal to counter their objections and reassure them that you will not let them down.

SITUATION 3. In your first departmental meeting, a colleague presents one of your ideas, which you'd discussed with her a couple of days before, as her own. How do you react?

Solution A. You say nothing because you're worried about causing an argument in front of everyone. But you decide to have a word with her afterwards to set the record straight.

Solution B. You express disbelief and firmly point out that this was your idea in the first place. You go on to say you resent being treated like this in such an underhand way.

Solution C. You diplomatically point out that it was something you and she had discussed, because you had found that this particular idea had worked well in your previous job. You mention how pleased you are that she's taken it on board so quickly. You invite her to work with you on the project.

SITUATION 4. In your first week, you have an urgent project to complete. You don't want to let the department down, so you ask your assistant to help you. He says he has an even more important assignment to complete for another person so he can't help.

Solution A. You try to bribe him to fit your work in – it's something you've successfully done in your last firm.

Solution B. You try pulling rank and say there's no way this deadline can be missed. He has got to stay late and do the work. And you mention that you've never had any difficulty with your previous staff.

Solution C. Explain about the urgency, and tell him the reasons why the work has to be finished today. You offer to negotiate on his behalf about the other work he will have to lay aside to help you. You thank him for his helpful attitude and tell him you will repay the favour.

2.3.2.4 Overcoming difficulties

Should you be faced with dilemmas at work, it might seem easier to postpone dealing with them. However, you should know that nothing ever gets better by being put off. Why is it so common for people to procrastinate when facing awkward situations? There are usually three reasons:

- fear of being ignored;
- fear of humiliation;
- fear of being rejected.

Here are some suggestions you could keep in mind when dealing with situations that require tact and diplomacy. First, **acknowledge that there is a problem**. If you check your emotions, body sensations and thoughts, you will be in control of yourself. That will assist you in taking control of the issue. **Communicate carefully**, clearly and positively. If appropriate get support from a colleague or a superior. **Be flexible** in your approach and review your goals – what outcome would be best, what are you realistically likely to achieve?

Don't procrastinate – act now to confront the challenge. A problem doesn't get any easier to deal with if it is ignored. When engaging the other party, **pay attention**, listen – without interrupting. Show that you understand how they feel as well as what they are saying.

Analyse the problem. It is crucial to differentiate between the **facts** (these sales figures are incorrect), **assumptions** (the calculations must have been prepared by junior members of staff), **generalities** (you never check that your facts are right) and **emotions** (how can I possibly trust you?). **Respond quickly**. If there is any action you can take immediately to make things better, do so. Focus on this rather than the cause of past grievances.

It is not necessary to take things personally. Do not give a flat no answer, and don't apportion blame. It is unwise to make promises you cannot keep. If possible, retain a sense of humour – laughter can lower the temperature considerably.

2.3.2.5 Never make an enemy where you could create an alliance

If you think you have lost the trust of a colleague for some reason, lose no time in trying to win back their confidence. It could be that someone feels slighted because they have not been consulted about a particular change in working procedures, or perhaps their views have not been noted about how things have been managed in the past.

> **Networking hint**
> If you think you are about to lose an argument with a member of your team, do everything in your power to repair the damaged relationship before it goes too far. Being able to recover a situation whereby no one loses face often leads to stronger and deeper relationships.

The lesson here is to give advance warning of any changes you wish to make. Ask for team workers' views and comments. It is far easier to open up a consultative phase and then, after appropriate consideration, bring about changes with reasoned arguments, than simply to ignore people and say, 'Do as I say.' Taking the temperature of the department is a good way to avoid breakdown in communications and relationships. Staff will feel reassured that their fears were unwarranted and will be more likely to trust your judgement in future.

2.3.3 Anger management

- Do you ever get involved in conflict?
- Does your voice sometimes develop a hard edge?
- Have you ever slammed the phone down on someone?
- Do you behave towards others in an adult fashion?
- Does your inner child sometimes escape?

There may well be someone at your office who has a tendency to 'throw their rattle out of the pram'. Networkers should try to adopt the techniques of a smooth operator. Solutions to angry exchanges involve using diffusing techniques. This is all about confidence and chemistry.

In the first place, stay cool. You can help or hinder a difficult situation by controlling your voice. Your voice should contain no hint of annoyance, arrogance or nastiness, only obvious concern and interest for the other party. This will instantly transform a battleground into a playground.

When someone is angry with you, move towards them verbally. Your manager may be incensed about something you have done. Meet him on equal terms. If you are not on the defensive, it will stop him being so aggressive. If he is raising his voice, lower yours. Use open questions when enquiring about the problem. Look for ways to resolve the issue by suggesting some possible solutions. If in difficulty, 'suspend reaction'. If you don't erupt with fury, or resort to sarcasm, he can't keep fanning the flames. It is very difficult to continue an argument if only one side is arguing.

When it is your turn to respond, stay calm. One angry person is quite enough. As a cool networker, be sympathetic. Show them that you understand and are anxious to deal with the problem. Tell them what you are prepared to do about it.

Allow people to express their anger. This is the best way to diffuse the situation. Keep your voice low-pitched, stay in control and take notes. If the situation requires it, take responsibility for sorting out the problem.

Pause before you react. Don't turn a crisis into a catastrophe. There's no need for panic. Urgent problems can often be solved quite quickly. But, first of all, get to the bottom of it. Find out exactly what is happening. You may not have had the full story from the person who first tells you about it. Ask questions.

Use your head. Have you ever been in a similar situation before? How did you handle it? If you have coped before, you can almost certainly do so again. The same solution may work this time, or you may have to get creative.

Is the situation going to need time? Create some space to deal with it. Cancel meetings so that you have some spare capacity. Don't add to your problems by failing to turn up somewhere you are expected. If there are other people involved who could make the situation even more difficult, deal with them quickly but firmly.

Relish the unexpected. Rise to the challenge of dealing with a problematic situation. In your old job you may not have had such an opportunity. See this as an exciting aspect of your new networking strategy and learn how to be a smooth operator. If you are completely stuck, think about someone you admire or respect – tap into your virtual team for help. Ask their advice, what would they do in the circumstances?

2.3.3.1 Dealing with criticism

Listening skills, like all communication skills, can, if used wisely, help diffuse the most awkward situations.

> **Networking hint**
> If you are being 'got at' by someone, it is best to keep your cool. If you can listen without showing any negative or defensive emotions, you will make things easier for yourself.

First of all, summarise the key points. Outline what the other person has said to make sure you've understood correctly. The more specific the criticism is, the more helpful it is. Find out, by asking questions, exactly what action has given rise to this particular situation. Was it the behaviour of one person? What impression have they formed and why was it unfavourable?

Criticism is rarely groundless, but, due to heightened emotions, can often be exaggerated. If you can swiftly extract the elements that are useful, you can turn them to positive advantage by acting differently in future. You will find the experience useful, if only to avoid a recurrence of the situation.

2.3.3.2 You can always try a bit of flattery

If you want to become an expert at building good relationships, ask those responsible for raising the criticism for their help. By seeking their advice and making them part of the solution strategy, they are likely to form a favourable impression of you. They could turn out to be a useful ally, influential as a mentor, coach or referrer, if handled correctly.

2.3.3.3 Always think positively

The people who have criticized you have alerted you to a number of things. They may not realize that not only have they given you free information but have enabled you to increase your knowledge of interdepartmental relationships. This will help you to improve your survival strategy and future planning. By implementing a solution, you have taken positive steps to avoid similar situations occurring and improve relationships with other work colleagues.

2.3.3.4 Wherever possible, give praise

Whether it is staff members who have helped you sort out the problem, or those who raised the criticism in the first place, take time to say thank you. By praising others for what they have done well or contributed, you will expertly disarm your critics and you will reinforce the message that your behaviour is exemplary under difficult circumstances.

2.3.3.5 Respect and trust

It is when trust waivers that relationships with colleagues become shaky. If they cannot rely on your word, you will not be able to rely on them. You cannot make claims about your professional ability if you don't believe them and deliver them. Nothing worth doing is achieved overnight. If you convince people about your trustworthiness, you need to reassure them that your motives are not self-centred. Self-preoccupation can wreck the start of many promising working relationships.

> **Networking hint**
> The establishment of trust and respect in relationships at work is paramount. Your ability to keep promises is what speaks volumes. Whether it's about keeping to time, returning calls, providing promised information, working to agreed budgets – whatever the issues – you need to be consistent and not let anyone down.

2.3.3.6 Don't put yourself ahead of your colleagues' or boss's needs

Take the time to find out who's who in your office hierarchy. You can work out where you stand, and where your colleagues and close teammates are in the command chain.

- How does your boss fit in?
- Who is his manager?
- Do they both get along?
- Can you do anything to help your boss impress his seniors?

Arrogance, ego and the need to prove yourself right will work against the situation. Your pursuit of the relationship should be for their benefit in the first place. When they believe this, you will gain from the relationship. Remember to be a 'giver', not a 'taker'.

Transparency is important, so make your objectives clear and understandable. I cannot stress enough the importance of clarity over ambiguity. If your intentions are not made clear your colleagues will not know how best to help you.

You may not necessarily be in a position of influence or able to provide powerful help. However, simple things, such as handing in a report you've been asked to produce a couple of days early, may make a difference. Your boss could show it to their boss ahead of schedule, which would put them in a good light.

Sharing information with them that possibly seems to you irrelevant or inconsequential could have far-reaching effects. Networkers have plenty of reason to listen to everyone (as a method of familiarization) and to read whatever comes their way (as a means of information absorption). So don't shy away from listening to chatter at the coffee machine or reading whatever is available. You never know what you will hear or learn.

Here is a checklist for those who want to develop relationships that work.

- Be transparent in your actions.
- Communicate with all sides as well as upwards and downwards.
- Network extensively to keep well informed.
- Identify and watch the 'politicians'.
- Put yourself in other people's shoes.
- Anticipate and manage others' reactions.

Chapter 3
Making connections

In the previous chapter, effective networking skills were covered. You've attended events and collected the names of potential contacts. Next question: how do you turn these into valuable business connections? And this is where we tackle the second stage of the three-stage plan we introduced above. It's time to set out your objectives, understand and document your goals.

Decide from those you've recently met, or those already in your database, with whom you want to build connections. Then you'll need to work out how to do it. Starting simply is always the best. Focus on success and the business development you will achieve.

If you are going to be maintaining a database on which you will rely heavily, you must ensure it is in a healthy state.

> **Networking hint**
> Begin by identifying common links and themes among those in your database. With practice, you can usually find two or three things in common within a few minutes. Use this as the common ground from which to build a valuable connection.

If you were a gardener and about to plant some new flowers, you would prepare the ground before you put in the bulbs. Now is an ideal opportunity to create some space. Remove any contacts who are out of date or incorrect. Add potentially new and exciting contacts to a database full of inaccuracies, and they will get swamped by the old material and will sink without trace within a short space of time.

Some of the information that you should be aiming to record about your new business connections as they develop are:

- contact's name, address, telephone number, fax and email;
- company name and job title;
- source of original meeting, venue or person who introduced;
- record of what transpired at first meeting;
- type of person/reaction – cold/tepid warm/hot;
- arrangement for follow-up – timing/method;
- personal details – birthday/family/hobbies and interests;
- geographical details – area of country if visiting them;

- background – them/their company/previous positions held;
- aims and objectives, links and shared acquaintances.

Spending some time on cleansing and updating your database not only refreshes your memory as to what's in it, but it helps you work out which contacts will connect well with your new acquaintances. From here you can begin to create even more exciting and harmonious relationships that you can develop with your newly acquired skills.

3.1 Attitudes and approaches

3.1.1 Responsiveness

One of the categories listed above – whether a person is cold/tepid or warm/hot – may seem slightly unusual to you. But it is something quite often used when measuring the level of people's responsiveness. There are four degrees of reaction when you meet someone for the first time.

- *Type 1: assertive – cold*

These people do not trust new contacts. They are introverted and do not welcome approaches from other people. They prefer to remain aloof. It will not be easy to penetrate their reserve.

Do not expect a warm welcome when you meet them. Accept their negative attitude. It is not personal. Use your professionalism as a foil. Keep small talk to an absolute minimum. Emphasise that you are talking to them for sound business reasons. Make your opening remarks short and very much to the point.

- *Type 2: accommodating – cold*

These people are a little warmer than Type 1. The best way is to let them take the lead. Demonstrate that you are in control of the interview by attentive listening, note taking and asking concise, factual and open questions. These will help to direct the meeting. Be firm and polite but never subservient. Position yourself as confident, professional and calmly determined.

- *Type 3: accommodating – warm*

You can expect a warm welcome, but so can everyone else. Their warmth does not indicate that you are particularly special. Allow them to express their feelings with some small talk but stay in control. Do not lose sight of the fact that you are there for business reasons.

These people like to think they belong to select groups, so mention as early as possible the involvement you/your company have with other comparable, reputable companies. Tell them how you would like to progress the relationship, including your role and theirs.

Keep the opening conversational and flexible. Position yourself as a friendly contact.

- *Type 4: assertive – warm*
Expect a correct and professional opening with a warm handshake. Be as professional as you can. This person will expect you to acknowledge them as commercially astute. Your opening remarks should be short and clearly indicate the purpose behind the meeting. Flexibility is the key so that their ideas can be accommodated in a joint desire for the business relationship to progress. Be prepared to review your objectives and avoid standard approaches or responses.

3.1.2 Review – a quarterly or half-yearly update

Check your closest contacts first to ensure that their details have not changed. If someone has been promoted, make sure that you have a note of their correct job title. Your valuable contact will not be pleased at being addressed as 'senior associate' when she has just been made up to director level. Any misinformation that remains uncorrected works against agreeable business relationships.

Make a note in your diary to update your contacts on a regular basis. If you do this carefully it won't become a mammoth task. This will lessen the risk of the job being put off to an indefinite date.

3.1.3 Check – who really is who?

At the same time, review the structure of your network of contacts. Are they categorized correctly so that it is possible to access people quickly? Do you have enough groups, categories and sub-sections? 'Client', 'prospect', 'supplier' is OK but it won't really be sufficiently detailed. You could include clubs, organizations, profession, industry, sector. Say you're attending a conference in Bristol next month. Can you find, at the press of a button, all the business contacts you have in that geographical area? Do you have time to organize visits to them or entertain them while you are there?

Create whatever fields you need so that you have the information accessible at the press of a button. Do you keep the notes section updated each time you make contact or meet?

> **Networking hint**
> When you have a systematic approach to keeping your contacts list neat, clean and tidy, you'll use it more often and effectively. Self-discipline and orderly procedures make it a valuable accessory.

3.1.4 A systematic approach

Set up a monthly reminder note to contact anyone you haven't seen or spoken to in the last six to eight weeks. They will appreciate your keeping in touch. A friendly enquiry as to how they are may be all that's required. Many people will be amazed that you've rung them without any particular reason or ulterior motive attached.

It's worth repeating something that's already been mentioned: building long-term, trusting and respectful relationships takes time, not only in personal matters but also in business. If you rush it you will be disappointed.

Developing connections with like-minded people with whom you can do business, either now or in the future, is the aim. One of the best ways is to try to help them as much as they can help you.

When working to create brilliant business connections, don't compartmentalize your contacts too rigidly. Links between people are unplanned and spontaneous. (It could be something as simple as discovering that you both worked for the same company some years previously.) Coincidences often occur and by avoiding boundaries and boxes you will be open to every opportunity as it presents itself.

If you put into practice the foregoing suggestions, you should be building and strengthening an effective and valuable collection of contacts, which is regularly refreshed and expanded.

Keep lines of communication clean and clear and use your network to develop new and exciting lines of approach. It is better to have a smaller and more manageable collection of contacts than something large, unwieldy and inaccurate.

3.2 Applied thinking

Who are you looking for and how do you put your contacts to work? Are you considering moving to another job within your company? How can your database help you?

For a start, it would be sensible to speak to people who are already working in that department or who have held a similar post elsewhere. As well as your peer group, consider talking to junior staff. They will have a different take on the organization which could be useful.

Look at the situation from all possible angles. Whom do you know who could give you advice or information? Think laterally as well as upwards and downwards. You could consider finding someone who would act as a possible mentor. They may be older and more experienced and have adopted a similar approach some years previously and guide you towards meeting your goals.

3.3 Categorizing your business connections

Why do you network? What activities do you pursue? Where do you do it?

Anyone serious about building corporate relationships has to be focused on what they want to achieve. Everyone in your database is a business connection. This includes existing clients, prospects, former clients, suppliers, influencers, bridges, links and gatekeepers. It can also include those with whom you work, former colleagues, past employees, ex-employers. Beyond that, it reaches members of your clubs, professional associations, associated businesses and other networks. It can also include your friends and acquaintances.

How you make use of these connections depends on what you want to achieve and what activities you pursue. Be selective. It's impossible to keep in touch with everybody you meet, and it's not necessary to do so.

There comes a time when people move on and you lose touch. But, with good organization and an effective network, you can maintain a link with these contacts through shared third parties. But one important factor in all this is motivation. If you lose interest in the process, everything falls apart.

If apathy prevails, nothing will happen. There has to be an incentive to continue to build your network, increase your contacts base and develop business relationships. This comes down to your own personal attitude.

3.4 Suggested methods of connecting

Some people host networking events, others belong to a selected number of professional associations and attend their meetings regularly. You can use a number of opportunities, such as receptions, parties, industry-related conferences, seminars and workshops. Attending exhibitions, professional interest-group workshops, private social functions and sport and leisure events are also useful.

There is limitless opportunity these days to meet people and foster good connections. But it's only with persistence that the relationships will flourish. You can't become 'best friends' with someone by putting in sporadic appearances every six months. Whatever you do, wherever you feel most comfortable, be alert and open to making connections.

In addition to your profession, the event at which you are both present, or the fact that you have been introduced through a mutual friend or a colleague, you may have other links in common.

> **Networking hint**
> The difference between 'networking' and 'connecting' is that there is more than one common thread running throughout.

3.5 Coincidences do occur

Recently, a friend of mine invited me to attend a reception. Unfortunately, on that particular evening, illness prevented my friend from attending, but he suggested I go along anyway. I did what a lot of people dread, and walked into a room full of strangers. The first person who spoke to me was (like everyone else in the room) totally unknown to me.

To break the ice, I asked him what he did. He said he'd retired last year from a career in teaching. I enquired as to where he'd taught and he named a few schools, one of which I recognized. I told him that a friend of mine was teaching there. He was amazed. My friend and this man had been colleagues for over 25 years and he'd been best man at his wedding.

To recap: **networking** is the essential **first** stage (as discussed in Chapter 2). It is the framework or skeleton. Making **connections** is the **second** step, putting flesh on the bones and developing the frame, as described in this chapter. The **third** and final stage, **building relationships**, is where you breathe life into the process. That is when the muscles on the framework are made to work (see Chapter 4).

3.6 New approaches

Invest some time in polishing up your existing network. Work out where your new contacts fit in, how they dovetail into your existing contacts. Spend time working out how best to categorize and detail them. Think about whether they are decision makers, influencers, bridges, links or gatekeepers.

- A **decision maker** is someone who can award contracts and has the power to agree to something you wish to happen.
- An **influencer**'s word carries weight if he mentions you to the decision maker – sometimes called a recommender.
- A **bridge** is someone who can introduce you to another person you want to contact whom you can't otherwise reach.
- A **link** is a shared connection between you and someone else that helps establish credibility and trust with the new person – not to be confused with name dropping.
- A **gatekeeper** is the person who stands between you and your desired contact – always be extra nice to them.

In today's world of ever-changing jobs and faster communication, it pays to be respectful and polite to everyone wherever possible. You never know where you will encounter them again. Someone you meet today could easily turn up in six months' time at another location or organization. If they remember you, maybe they could influence someone favourably on your behalf.

Decide who among your contacts belongs in which category in your database. Do you know where the gaps are and how you are going to set about filling them? Who can help you to do this and what can you offer them in return?

To assist you – here are a few 'A' words to consider:

- How should you **approach** people?
- What is your **aim**? To communicate **appropriately**.
- **Ask questions** – enquire first and then listen.
- Develop an **attitude** of gratitude.
- What **action** should you take to progress the connection?
- Offer **assistance** and **ask** for help.
- When enquiring about something be **attentive** and **aware**.
- Keep **alert** to the possibilities of **alliances** which can **add** value.
- Learn to **appreciate** others and their **abilities**.
- **Analyse** situations to your own **advantage**.
- Be **adaptable** to other people's needs.

3.7 The case for developing business connections

There is no doubt, close connections in business save time and money. Business effectiveness depends as much on human-related activities – relationships, interpersonal skills and communication – as on technical skills and abilities. So how do you identify the best business connections?

Anyone is potentially a good connector. But they must understand what you offer or need. If you don't take the time to explain this to them, how can they possibly help you, or vice versa? This is why, when you encounter someone with whom you feel instant rapport, make sure you get a half-hour meeting with them shortly afterwards, to explore whether this was mutual.

You need to spend time with a new contact fairly early on in the process, to establish how best you can help each other. It could be that one of you is new to the company you work for, so could want information on who's who and how things are done.

Alternatively, you could be **looking for information or researching** a particular subject about which this person is an expert. If so, consider ways in which you can be of use to them.

People who are skilled at developing business connections integrate the process into their lives. They have wide-ranging contacts with whom they keep in constant touch. They exchange ideas and information and offer help. You probably recall meeting such people. Maybe you are one of them.

Their outstanding characteristics include sincerity, curiosity, consideration, sharing, understanding and appreciation. They are the **givers rather than takers**. They keep in touch with their contacts even though there is seemingly 'nothing in it for them'. They are always open to opportunities to assist people and to broaden their network by sharing connections.

3.8 Reasons to get started

If you want to turn your contacts into connections, you need to work at cementing relationships. How is this done? Simply by getting to know people better. Make the time and opportunity for face-to-face meetings whenever possible. Early on in the process, one-to-one meetings are best. They help to build rapport quickly and easily. Further on in the relationship it is often helpful to include other interested parties.

> **Networking hint**
> The responsibility for getting communication right lies firmly with the communicator.

Clarity and appropriateness are important. Checking your intentions is also vital. Ask yourself these questions.

- What does your new business contact need to know?
- How much detail should you give them?

- What sort of action do you want to have as a result of this exchange?
- How is it to be done?
- How do you want them to feel – in agreement, pleased, enthusiastic?

If you are clear as to why something is being done, and you have set yourself defined objectives, then plan how you are going to deliver the message. Is significant feedback necessary? If so, it should be handled by a face-to-face meeting. Is simple feedback necessary? Possibly a phone conversation will suffice. Does complexity require a combination of both methods?

In each situation a number of different factors will influence the chosen method: urgency, complexity, formality, involvement of a number of other parties, etc. Every possible method of approach needs to be considered. Choice should be made with regard to such criteria, and avoid the temptation to do what is easiest.

3.9 Planning the meeting

If you choose a face-to-face meeting, remember they may not be everyone's favourite activity.

Getting your face-to-face meeting off to a good start involves responsibility. If you are being proactive in creating a close relationship with someone who could be vitally important to your career or business-development plans, remember these points:

- make the meeting positive;
- ensure its purpose is clear;
- establish your authority and engage the other person;
- create the right atmosphere – friendly and flexible;
- generate interest and enthusiasm for the process;
- be professional and businesslike but informal;
- keep to time limits – respect other people's schedules.

A well-run meeting demands concentration; you should avoid interruptions and make the process effective in the following ways:

- actively stimulate creative thinking;
- contribute new ideas and encourage others to do the same;
- steer the discussion into new or unusual directions;
- find new ways of looking at things;
- consider novel approaches and give them a chance;
- aim to overcome obstacles to progress.

Communication is a two-way process – more of this later. If you want to stimulate feedback, comment, exchange of ideas and suggestions, you should positively encourage it. To ensure your business relationship starts well, encourage communication with your new contact:

- create a situation where two-way communication is expected;
- stimulate it by regularly keeping in touch by a variety of means;
- make it easy – provide feedback and useful pieces of information;

- react positively to each and every exchange – acknowledge and thank your contact;
- give credit – this will ensure a flow of ideas and exchanges;
- make time to do this on a regular basis, however busy you are;
- be available when your business contact makes an approach.

Finally, keep the possibilities for creating a proactive relationship in mind. Encourage the process, be versatile and flexible to keep your contact's attention. For example, don't just 'do lunch'. Invite your new contact to a variety of different events. A business breakfast is a good way of keeping in touch and not spending a lot of time 'catching up'. Much progress can be made in an hour before the working day starts.

If your contact is an evening person, rather than an early starter, meet for a drink after work. A 'happy hour' approach is informal and relaxed and can work wonders if that is a comfortable way of connecting for both of you.

Whichever method of approach you use, make sure it is a continual process. Set aside a proportion of your working week for relationship building and you will be amazed at the results.

Following up a face-to-face meeting can often best be done in writing. If this is what you propose to do, try to be:

- concise (as brief as content and purpose allow);
- understandable (avoiding ambiguity);
- precise (say what is necessary and no more);
- jargon-free – use plain language (without technical vocabulary or terms);
- simple (in language and grammatical structure);
- descriptive (letting your words add to the message).

When there's no 'reason' to meet, keep in touch remotely. You can phone, email or write. Use the opportunity to carry out some research, gather market knowledge, exchange ideas or share contacts. While you are being generous with your time and connections, your business contact should be busy trying to reciprocate. You'll earn respect and your reputation will be enhanced. In addition, you'll make more new friends and have a good time.

Natural business connectors have certain things in common:

- they treat everyone as being interesting, special and likeable;
- they use good eye contact and positive body language;
- they make other people, particularly new **acquaintances**, feel safe and part of the occasion;
- they introduce people to each other effortlessly, remembering names and something relevant about those they introduce.

In a word, they have charisma.

3.10 Go on a charm offensive

Have you met someone like this? When you were introduced, did they smile, enter into conversation easily and draw you out? They probably asked you questions about yourself, and listened to what you said. They made you feel important. When you parted, you probably thought to yourself, 'What a great person!' Not only is it easy to be in their company, they are at ease with themselves.

Does this describe someone you know, or the person you want to be?

Charm is contagious. To be a charmer, you should be assertive but not arrogant, vain or conceited. Charmers are confident and they build self-esteem in others. They have broad horizons, are not narrow-minded and are enthusiastic. Above all, they are curious, asking lots of questions. They empathize, are responsive, have a great sense of humour and are frequently self-deprecating. They are at ease using a number of different communication styles and often employ 'mirroring' techniques. And, they know how to listen.

3.11 Mastering the art of good questions

Good questions are perhaps the single most powerful interpersonal tool in building business connections. A good question is one that shows you are interested in the other person. It's an open question – requiring more than just a yes-or-no answer. The more questions you ask, the more you are encouraging that person to trust you.

> **Networking hint**
> Good questioning builds great relationships. It indicates that you care about the other person and their situation.

Good questions convey competence, sincerity and sympathy. You should practise the art of listening, too. By paying attention to what people say, you can note this information for future use. If you greet someone the next time you meet by asking them how they enjoyed their holiday in Antigua (assuming that it was their holiday destination!) you'll rise so high in their estimation.

3.12 Seven steps for making connections

You've networked to make connections. You now recognize the crucial differences between networking, connecting and relationship building.

3.12.1 Step 1: Database

Hopefully, you've designed and customised your business network architecture (the database). The next step is to audit existing and past connections. You have noted the relationships that are still valuable or potentially useful, and added all the new information you've collected. Have you designated each contact an appropriate category – cross-referencing where necessary?

3.12.2 Step 2: Categories

Now you are looking to link your contacts together. Start by connecting individuals, where there are possible collaborations, complementary skills and services, potential partnerships and alliances.

At this point you are in a position to help your contacts by sharing connections to win, retain and develop their careers or business. Internally, within your company, you're aware of the importance of relationship building, empowering individuals and teams. You're ready to facilitate connections between employees, employers and other influencers, to deliver outstanding results.

3.12.3 Step 3: Links

With practice you'll be aware of the skills required to handle delicate issues, where sensitive business connections and confidentiality are paramount. You will also be noting which people would be best positioned to assist you when these opportunities arise.

3.12.4 Step 4: Keeping in touch

It always pays dividends to let your contacts know that you consider them special and worthy of particular value. That alone can set you apart from others, particularly those who do not understand the benefits of one-to-one attention.

Don't forget that, when one of your contacts leaves a company, you have diverging paths. Not only do you want to keep in touch with your friend in his new position, but you will also want to start a rapport with his successor in post. So many people fail to track their contacts when they relocate and risk losing a valuable resource in the process.

By establishing and maintaining communication with your contacts on a regular basis, you are putting yourself ahead of many people. You will earn respect, encourage trust and add value to the relationship you are nurturing and developing. Keep your connections warm and friendly. People naturally gravitate towards those who they feel will be empathetic. Pay attention to your contacts' needs and wants, and, when the time comes, let them know how you can help.

3.12.5 Step 5: Keeping notes

If you make a habit of detailing information, it will pay off in spades. This means keeping notes of meetings, exchanges, telephone conversations. Within a few months you'll most likely be able to offer something relevant to their needs. No matter how tenuous the link, your contacts will register the fact that you remembered them. Irrespective of whether the moment has passed and they no longer need that contact or service, it will emphasise to them, 'Yes, he really was listening.'

3.12.6 Step 6: Being distinctive

It's not hugely important what you do to make yourself 'stand apart' from the pack as long as you do it. What works best is something that is affordable, attainable and relatively easy to offer. If you have access to information that could be advantageous to your contacts, offer it to them. You may have some sporting interests in common, or facilities that they could use, in which case invite them to participate. Whatever method you employ, to be distinguishable in business is a great form of advertising. It speaks volumes not only for you personally, but also your company.

3.12.7 Step 7: Proactivity

By being proactive in engaging with your contacts, you will be setting yourself and your company apart from its competitors. Remembering people's names and personal details may seem unimportant, but even the smallest nugget of information can keep you in the forefront of a contact's mind.

Establishing yourself as an 'identity' means that your contacts will remember you ahead of other people they have in their database. Be outstanding, and they'll go out of their way to help you and your business.

3.13 Relationship building

This is where we touch on the third stage in the three-stage plan we talked of above, and we'll deal more fully with relationships in Chapter 4. So how *do* you build deep, meaningful business relationships? There's no easy answer to this; neither is there a quick fix. There are a few important points to bear in mind before going on to develop the best strategy to suit your own purposes.

3.13.1 A two-way process

Any business relationship is a two-way process. The person with whom you are dealing will want to:

- feel important and to be respected;
- have her needs considered;
- be able to trust you;
- want your input and ideas to help her;
- be advised in advance of any snags/problems.

This may sound a bit of a tall order. But, until you have convinced your new business contact that you are professional, reliable, discreet and honest, the relationship will not proceed far. You need to develop some self-awareness too. Self-awareness enables you to be genuine.

It may be impossible ever to know yourself completely, but, if you are trying to build good relationships with people, being aware of your own personal strengths, weaknesses, hang-ups and prejudices is helpful to the process. What this means is your attitudes and values. If you are aware of these, you will then know how to adapt your own traits so that the relationship building is not inadvertently hindered by your own behaviour.

Interacting between parties requires perception and sensitivity. Effective relationship building is achieved by being alert in mind, emotion, body and spirit. Concentration levels need to be high. You are starting a process of offering something to others in the hope of obtaining something in return. Confidence is a quality that springs from competence. People show confidence in different ways. Some are robust and self-assured. Others may be risk takers and not daunted by challenging situations.

Keep an open mind. Every meeting with your business contacts will be unique and present different opportunities. If you are resourceful and adaptable, you won't need to have ready answers. What you are aiming for is to appear trustworthy and be able to trust others in return.

3.13.2 Watch your step

The relationship-building process is like dancing. Each stage needs to be taken willingly and in tandem with the other party, before you move on to the next one. Sometimes a pause will be required in between, in order to assess progress and regroup. The process needs to be balanced: each party should be open in negotiations, able to weigh up the pros and cons of the information exchanged, before making a decision to take a further step.

> **Networking hint**
> No relationship will be perfect; neither will it be easy or quick to achieve. In fact, the old saying 'more haste, less speed' is very apt.

As long as a positive balance is maintained, progress will be made. The process can sometimes be quite complex, depending upon the size and nature of your contact's business. If the plan (on your part) is to move closer to an organization where a major project is likely to be awarded, then the success or failure of the relationship you are attempting to build can be dramatically affected by a number of factors.

This could be other parties involved in the decision-making process, or competitors of yours who may also be lining up for favours. As long as you remember that this process is inherently two-way, and that satisfaction is needed on both sides for it to continue, then you will be able to make progress towards a winning outcome.

In any complex process where there is a need to relate closely to someone, this can be done best if it is thought through carefully and approached in a planned and structured way. Effectively, you are trying to play an influential part in the other side's decision-making process.

Your objective is to assist them to make decisions – the right ones – that will have the result of giving you the desired outcome. The more you are prepared to 'give', the more likely you are to achieve your goal.

3.13.3 Keep on track

Your plan (structured approach) needs to be like a map. Even if you have to deviate from the planned route, the chart should help you keep as close to your track as possible. The overall process of relationship building is multifaceted. It needs to be controlled and managed. You will need to use ploys and gambits, as well as playing to your own unique strengths, to nurture the process. Try to keep in mind the importance of seeing things from the other side's point of view. Fine-tuning is paramount; being able to deploy relevant approaches from all your available interpersonal skills is advantageous.

3.13.4 A learning curve

Learn something from every encounter, even if it's how not to do it in future. If you are embarking on building brilliant business connections, you will probably be making hundreds of business-development calls to your contacts every year, and arranging lots of meetings.

One way to maximize the amount of the time you've spent doing this is to ask for feedback. The amount of feedback available from these contacts is massive. Do you take advantage of this? Review your last six meetings or contact-building calls. Did you ask yourself the following?

- Why did they say what they did?
- Did they misunderstand something I said?
- Why did/didn't they agree?
- What did they really mean?

Try to get into the habit of reviewing each exchange in your mind after you've finished, and analyse the results to improve your relationship building the next time. By adjusting the way you communicate with contacts you will avoid repeating your mistakes. Your overtures will appear fresh and well directed in future.

> **Networking hint**
> In business dealings, always prepare. Good planning is one way to give yourself a head start over any competitors.

Preparing yourself for your meeting with an important prospect could mean just taking a few minutes of your time thinking matters through before you pick up the phone or start a meeting. Or perhaps you should sit round a table with a couple of colleagues discussing the best way to approach a key prospect. Whatever method you use, make sure you always do it. Forewarned is forearmed.

If you are visiting someone who is very busy, the less of his time you take up will earn you points. Try to make the preamble fairly short – don't spend ten minutes explaining to him in minute detail what the traffic conditions were like on your journey. On the other hand, if a 'warm-up' session is necessary to put your business contact at ease, spend a few minutes asking him about one of his hobbies or a recent trip he's taken. Be sensitive to his reactions.

3.13.5 Set objectives

If you don't know where you are going, how will you know when you get there? One way of looking at the relationship building process is to regard it as starting out on a journey.

Each exchange with every prospect needs clear objectives.

- Do you want to win new business?
- Do you want them to act as influencers or referrers for your company?
- Are you trying to obtain market information?
- Are you hoping to work for their organization at some point?
- Do you want an introduction to a key decision maker?

If you don't have a clear idea of what you are trying to achieve, it is difficult to set in motion the actions you need to achieve it. Remember that well-known mnemonic – attributable to so many business activities. When you are trying to build brilliant business connections, remember to work **SMART:**

S	**Specific**	Be clear about what you want to achieve.
M	**Measurable**	Identify the stages so you can track progress.
A	**Achievable**	Can you really do it? Be honest. Don't be over-ambitious.
R	**Realistic**	Should you actually start the relationship-building process.
		Is this the right time for you and your company?
T	**Timed**	Work out the timing. When do you expect to meet your objectives? In weeks, or months?

3.13.6 Make it a habit

Balancing existing business relationships with finding new ones – which you need to grow and nurture – is a constant dilemma for anyone who wishes to progress. Some people simply find it almost impossible to fit in the time required among other work priorities. To do justice to the importance of it, you must adopt a systematic and consistent approach.

What you are trying to avoid here is a feast-and-famine scenario in your business: either having too much to do or not enough. Business connections used wisely should alleviate the possibilities of this. Another factor most people dread is 'cold prospecting'. If you nurture your contacts you should not have to tackle cold leads. All your connections – even the newest you are following – should be warm, if not actually hot.

If your plan is to provide a regular supply of interesting new projects, then the activity required to produce this stream of work should be carried out regularly and proactively.

Make sure it is strategic and fits with your company's business-development plans. New contacts are the lifeblood of most businesses, so make your first rule to be continually on the lookout for new opportunities and connections; be systematic and persistent.

3.13.7 Creating a method

Whether you are looking for new business, or information, or researching ideas, using one business contact to lead to another is a great way to continue to increase your sphere of influence. One way of making this work is to ask your business contacts who else they would recommend you to talk to about the matter under discussion. You can ask them if you can mention their name. Provided you offer them something in return, they will most likely be generous in giving you the information.

Ask them who among your contacts it would be useful for them to meet, and make a point of following up on that request. Send them the information, invite them to the appropriate function, fix the next meeting, or introduce them to the third party at the next opportunity.

3.13.8 Centres of influence

There are people and organizations with whom you have made contact on a regular basis who have the power to influence or recommend. These could be trade and professional bodies, chambers of commerce, associations, financial institutions and other professional service firms. By offering them some help or service, you will ensure they are reminded of what you do and are up to date on your company's activities.

Make a note of any business contact you have who could lead you to a number of other influential business people. If this reciprocal relationship is successful, it is a most cost-effective and worthwhile way of increasing your network. You can always win friends and influence people by offering to do some *'pro bono'* work. This could be volunteering to give some of your professional expertise to a favourite charity or not-for-profit organization; or to help raise funds for a specific project.

> **Networking hint**
> Anything that emphasises your willingness to be a 'giver' rather than a 'taker' will get you noticed and remembered.

3.13.9 Chance contacts

Create opportunities for yourself by being observant. This sounds obvious but is often easily overlooked. Scan your local paper, trade press, journals or professional publications. If you've recently read something about any of your business contacts, get in touch with them and say how pleased you were to see them mentioned in the news.

It could be that one of your contacts has just raised some money for a local hospital or charity. Maybe they, or their company, have won an award or a significant piece of new business.

They will be flattered and remember that you took the time to call them. By being observant and showing them that you take an interest in your contacts, they will remember you.

Don't ignore office gossip. You never know what you might learn. Maybe someone you know has moved into new premises across the road, or to the office suite on the floor above you. If they are on your contacts list, make a point of calling in to say hello. Anything you can do to establish yourself in their mind will be helpful when you are trying to develop your business relationships.

3.13.10   Expect the unexpected

You never know when a chance encounter will provide a connection. It doesn't always lead directly to work, but it could be useful in some way or other. It could be as simple as being able to remark to one of your business contacts, 'You'll never guess who I met the other day. He said he's a great friend of yours . . .'

3.13.11   When is the best time?

If you think you should make contact with people only during working hours, you may be putting yourself at a disadvantage. Many professionals these days are available round the clock. It is worth being flexible enough to be available at times that suit the needs of your business contacts. If you show that you're prepared to fit in with their timetable, it makes a lot of difference and can give you an advantage over others who do not.

3.13.12   What to say

When it comes to making the call, or attending the meeting, with your business contact, be sure that you know your own business, and also theirs.

- Keep your checklist handy if you're unsure.
- Awareness of the objectives of your call or meeting is crucial to success.
- Have your key points ready for opening the dialogue.
- Be prepared to handle objections or overcome obstacles.
- Make a note of any actions or decisions that need to be made as a result of your discussions.
- Finally, emphasise the importance of your meeting by following up afterwards, supplying information or offering services.

Think long-term

Good customer/client relationships are built over time and are therefore difficult to dislodge. If a relationship is founded on secure foundations it will endure the hiccups and glitches that are inevitable in commercial transactions. New best friends – started on the instant – may be great for the moment. But be warned, they are frequently transitory. If they are built on shifting sand, they can disappear just as rapidly as they appeared.

Being pro-active with your business contacts is essential while maintaining a positive mental attitude. Be confident, polite and persistent. If your industry or sector is prone to quiet times, it pays to stay in touch. Whatever your business or profession, it's not possible to be continuously engaged in a working relationship. By keeping in contact when you are not engaged on a project, you will be the one who is remembered when things improve.

Make it personal

You will get to know a great deal about someone over a period of time. Remembering a child's name or sending them a birthday card may seem trivial, but it will register that you are interested in them, not just the position they hold.

Think of your business contacts when you meet new people. Introducing your contact to people who matter in their industry or profession will make them feel important. They will know that they are part of your inner circle of business connections. In turn they will be conscious that they owe you a favour in return.

Listen to what they say. You can pick up clues about their hopes and fears. You might hear things about their organization or industry from an unlikely source. Passing on news and pieces of information may help them avoid upsets and unwelcome surprises. They will appreciate the fact that you gave good advice or warned them of impending changes and will be keen to help you in return.

3.13.13   Persistence pays

To harness the power of personal connections you need to keep a few 'P' words in mind.

- **Persistence** pays, there's no doubt about that.
- Building relationships is like **planting** seeds – they take time to germinate.
- One of the most important factors in the process is **preparation**.
- You need to prepare the ground – it pays to know about your **prospects**.
- You have to **persevere** – sometimes for months, and in some cases years.
- Try to be unfailingly **polite** and **patient**.
- A **positive** mental attitude and outlook is infectious.
- **Persuasion** tactics get easier with **practice**.
- Make sure you do some **planning** – so you know when and how to keep in contact.
- Don't underestimate the value of **praise** when communicating with your contacts.

3.13.14   Say it with feeling

What's important is the way you begin to build business relationships. In effect you're starting a process of persuasion. It's not easy, and often using words is just not enough.

You have to be able to hook the other party into the idea that there is something in it for them. Once they accept this, you are more likely to get their attention.

To be persuasive you should offer people reasons that reflect their point of view. You won't get far just saying why you think they should do something. Benefits are things that do something for people. The benefits of reading this book include helping you with your strategy for building valuable business relationships. If you are a cool professional type, make sure you communicate in a factual and efficient way. If you want to sound friendly, more informal and approachable, make your characteristics match the message you wish to convey.

3.13.15 Touching emotions and intellect

It is said that people act on emotion, and justify with logic. To be an effective persuader you should not only offer good reasons for something but also create emotional goodwill at the same time. If you need to persuade powerfully, bring in stories to connect with people's hearts as well as minds. If your objective was to motivate people to donate blood, you could tell a story about someone who needs blood for a serious operation. Successful fundraisers use emotive illustrations when persuading people to donate to charities, and their appeals can bring in thousands of pounds in revenue.

> **Networking hint**
> Recommendations work wonders. If you have shared connections, it is much easier to persuade when there are credible people to testify that your skills helped them in some way or other.

You should be able to come up with a number of people from your contacts whose name would add credibility and respect to yours. Do have the courtesy to check with them beforehand that you would like to mention their name to your new contact, as long as they have no objection. More often than not, there will be no problem. In fact, they may even offer to contact them in advance.

If possible, don't rely on one source for recommendations. Using several different parties gives further weight to your case. You increase your chances that one or other of your sources will be a powerful influence over the person with whom you're building up trust.

Ensuring that you are persuasive needs some preparation. Think about how you want to come across to your business contact. Ask yourself why anyone should want to listen to you.

List your reasons and then organize them.

- What are the most important things you are trying to say?
- How can you build rapport with one another?
- Can you arrange your thoughts into a logical sequence?
- You could start with something attention-grabbing and continue to maintain interest throughout the exchange.

- Perhaps you want to build up your case throughout the dialogue and end with some weighty fact that has masses of impact.

Whatever you want to do, prepare for your encounter with your business connection as if you were planning a presentation. Make your content understandable and attractive.

Chapter 4
Where to make a start

Now we look in more detail at the third stage of the three-stage plan: relationships. Let's consider for a moment developing relationships internally. Within your organization there are probably a number of different types of people. Some of these are likely to be easier to get on with than others. There will be some people you work with whom you may never want to talk to, let alone develop rapport with. Despite inherent difficulties, it helps if you bear in mind a few basic rules that allow you to deal with different types of people effectively.

4.1 The ins and outs of internal workplace relationships

Do working environments bring out the worst in people's characters in some cases? The answer is probably yes. Why should this be? Because so often at work, people are put under immense pressure to perform to the best of their best ability. There can be tight deadlines. They can be expected to produce pieces of work that are not purely a result of their own efforts. They may be in a situation – through no fault of their own – whereby they are heavily reliant on other people.

If those on whom they are dependent for information let them down, there is a lot at stake. Instead of being reasonable and easygoing, they can undergo swift character changes. Failing to produce a result can have dire consequences – for example, serious financial issues or career implications. As a result, of course, workplace relationships can and do suffer.

Acknowledging this fact is a start, but it needs to be taken further to make things easier. For someone who, on a daily basis, has to confront different types of character with wide-ranging behaviour patterns, it can be a physically draining and exhausting business. This is particularly so if you find such things do not come naturally to you. Maintaining the smooth running of a corporate environment is a complex process, even for the most experienced people. If, individually, workers are able to practise a number of core skills to help them avoid personality clashes with their colleagues, their work surroundings should be less fraught. The workplace may still fall way short of an oasis of tranquillity but at least it won't increase your stress level too much. Adopting an appropriate attitude is important. If you go to work each day anticipating a battle of wills with your colleagues, that is what you will probably get. Adopt the conciliatory approach and you may find the atmosphere altogether more pleasant.

4.2 Colourful characters

Most individuals come under pressure from forceful personalities at some time or other. Coping with competition and dealing with challenging and diverse characters in the workplace does no one any harm. In fact in many cases it is a good thing – a form of positive stress.

> **Networking hint**
> No one progresses in an environment of apathy or complacency. Stress frequently stimulates performance.

This is a tactic many employers resort to, in order to work out who can cope with challenges and who can't. There is a proverb about a tree that was unyielding and one that bent in the wind. If you can't stretch or be flexible, you will break.

If, for example, you are entering a company at the first tier of the management structure, it is likely that the directors or partners in that company will put pressure on you as a young manager. You should be able to show them how successfully you deal with a workload that, at times, threatens to engulf you. If you are able to do this, it will give a clear indication how far you are capable of rising.

In addition to dealing with the stress of a heavy workload, it is important to show you can also adeptly cope with a number of staffing issues. By maintaining a healthy atmosphere in your department, you will show that you have the ability to cope with workplace relationships. You could well be on track to progress higher up the company structure in a shorter time than others who are similarly placed. Senior partners and directors are conscious of the importance of succession planning and you would do well to bear this in mind.

It is all too easy under certain circumstances to become a 'difficult' person. You could feel that the environment in which you are working is hostile and unreasonable. Some people are unable to restrain their tempers and react angrily when provoked by others. If you feel this could happen to you, it is best to avoid confrontation as your reputation could be adversely affected. Your experience, even if it is not huge, will tell you that being positive, keeping up to date with your work, extending and improving your skills and thinking before acting makes you an effective person. It will also show others that you are willing and able to maintain appropriate workplace relationships.

4.3 Attitude counts

This is why attitude is so important. If you can learn to adapt your way of behaving by changing your attitude towards situations and colleagues, things should remain tolerable. You can't change people's personalities but you can adapt to suit the particular circumstances. Taking such action avoids friction, and negative internal relationships can become positive ones. Harmonious relationships do occur, but human beings will always have trouble in dealing with one another.

Nowhere is this more prevalent than at work. Maybe you have a job in a place full of conscientious workers. These people can often be underestimated, undervalued and underpaid, despite carrying out their duties competently. It is possible that there are a few with whom you work who are paid extremely generously for doing a lot less. What makes life problematic is the way they achieve this, even profiting from other people's efforts. This imbalance brings out a lot of resentment and jealousy among co-workers. Tempers fray and rows flourish.

So how do you develop smooth relationships with difficult people so you can get on with your job? What are the secrets of getting different people to perform well despite their diverse personalities? Assertiveness helps a good deal. Adopting an assertive attitude enables you to deal with tricky people and situations as and when they occur. When circumstances require it, by asserting yourself so that you do not feel undervalued or overruled, you will develop your own strengths and personality at the same time.

Companies cannot afford to stand still. Employees are constantly being encouraged to acquire new skills. It could happen that your company gets taken over and redundancies are on the cards. You may get a new management structure, be transferred to another department or be asked to do your job in a different way. People who are taken unawares in such situations are at a disadvantage compared with those who are capable of embracing change. Standing out from the crowd is the best way to get your next job, or promotion. Give yourself a head start by being able to assert yourself and be respected by others. Doing so in a positive, rather than a negative, way is by far the better method.

The way you relate to other people – family, friends and colleagues – reveals a lot about how successful you are likely to be at building good professional relationships. In the workplace if you use these skills it can impact negatively or positively on people who can influence your career. You may be able to act exactly as you want to at home (behave badly), but at work your behaviour can affect your career progression. It is important to learn how to use it to exert maximum effect. Should you be someone who is naturally popular, this is a bonus. If you have the ability to make other people feel good about themselves, you'll have a huge advantage. The most successful working relationships are built around mutual trust and respect.

Networking hint
Respect for others can be as simple as saying good morning and goodbye to your colleagues when you arrive and leave. People like being treated as if they matter.

If you are sincere, show concern for their views and listen to what they say, they will respond positively towards you. Don't forget to thank them when they communicate information. If you occasionally need to convey criticism, be particularly careful to ensure that it is offered in a constructive and positive way. The sandwich method, saying two positive things with one adverse thing in between, is one method.

4.4 Understanding the characters

Most people you work with will be reasonable, so developing rapport with them isn't a problem. But there are a few who have certain personality traits that you should bear in mind. There may be some you encounter with whom you may not be able to build positive relationships. Let's now look at the three main ones.

4.4.1 Destructive characters

These people live to dominate, insult and wound others. They have an openly hostile attitude towards colleagues and can be sarcastic, uncooperative and arrogant. Their know-it-all attitude makes them extremely hard to get on with. They won't hesitate to stab you in the back, if it serves their purpose. Because they cannot be trusted, you must tread carefully when you are dealing with them. They are usually self-appointed experts who, because of their huge egos, can never be wrong. They know everything and refuse to believe that anyone else has anything useful to say or contribute to the workplace.

If you had to choose adjectives that described their character qualities, these could include:

- bad-tempered;
- unkind;
- bossy;
- predatory;
- ferocious;
- obstinate.

They can (at worst) resemble wild beasts, safe only behind bars. They like being regarded as dangerous. It gives them a feeling of power that is undeserved. Really powerful people don't have to behave aggressively.

4.4.2 Disgruntled characters

Every organization has a few of these – whiners, moaners and grouches. They display all negative qualities and are fully paid-up members of the 'half-empty-glass club'. Life constantly shows them nothing but bad luck and they are happiest when they are miserable. They are expert nitpickers. The last thing they want is someone with a positive attitude around them.

Solutions never feature on their horizon – they enjoy wallowing in their problems too much. They simply hate colleagues coming along showing enthusiasm for challenges. Their main reason for existing is for dampening down positive people, chucking cold water over any beneficial schemes and suggestions. Unfortunately, their attitude is infectious and can spread like a bad dose of flu to other members of the workforce. Adjectives that describe them best include:

- dissatisfied;
- frustrated;

Where to make a start   57

- sulky;
- complaining;
- envious.

Colleagues refer to them as moody, crabby and carping. They refuse to enthuse. You could offer them a couple of alternatives, and let them choose their perceived 'lesser evil'. Then work on that method to get the desired outcome.

4.4.3   Deadpan characters

These people are hard work. They don't have much courage and hardly ever say anything about what they think or feel. They would rather remain silent, or hidden from view. They'll do whatever it takes to avoid facing issues head on. Such people limit their communication to the occasional murmur. On a good day you might get a monosyllabic answer. They are expert at concealing their feelings and frequently have a mysterious air about them. If they were unwell, it would be hard to find out what their illness was.

They talk in cryptic fashion, in hush-hush tones, and prefer to remain anonymous, behind the scenes. They veil their expressions, their remarks and their behaviour patterns resemble something clandestine, undercover, stifled and suppressed. It is not easy getting them to open up, so most people give up and move on. Sometimes it is a case of 'the lights are on but nobody's home'. The adjectives that could describe them include:

- elusive;
- solitary;
- timid;
- secretive;
- remote.

Do these people ever open up? Their self-image is low or virtually non-existent, so it takes a long time.

4.5   Weapons of mass destruction

So what ammunition do you need to keep yourself intact when confronted with such characters? The trick is to appear indestructible, but how can you do this? Here are three suggested methods, briefly described. They are explained in greater detail later.

- **Verbal**: Use words to keep any communication under your control to defend yourself or support your point of view. If you have good conversational ability, you can use words skilfully to make them your ammunition against the enemy.
- **Negotiation**: By learning the art of compromise, or deal brokering, you can save yourself huge amounts of trouble. You should, with practice, be able to devise win–win strategies to get something out of any workplace relationship.

- **Humour**: This is an essential aid to defusing tricky issues. By reducing the temperature of a heated encounter, tempers can be cooled. Keep the humour directed towards yourself, and you may even emerge from the skirmish unscathed and the relationship may survive to be continued at another time.

4.6 How difficult are you?

Enough said for the moment about 'the others'. What about *your* personality? When other people are being difficult, do you mirror their behaviour? As mentioned before, you have a responsibility not only to yourself but also to your colleagues. If you are finding your work difficult and you're under pressure, you could be giving off negative signals to others.

Remember that your personality is made up of four main forces:

- emotional needs;
- economic needs;
- models;
- values.

People who were nurtured from an early age in a loving environment are more likely to demonstrate affection and be kind to others. On the same basis, those whose upbringing was materially comfortable are probably naturally generous and don't possess miserly tendencies. The role models you choose in your early life influence the way you behave, just as your values are formed by your education, environment and job.

4.6.1 Self-knowledge helps

To be able to deal successfully with other people, the better you know yourself, the easier you will find dealing with others to be. Some of the individuals you find challenging may not really be so. It may simply be the fact that you are not used to them and feel intimidated by them. They could be more secure and confident than you. Ask one or two other people whose opinions you trust what they think of them. They may say that your fears are unfounded.

> **Networking hint**
> To have a reasonable discussion with a demanding person needs a level-playing-field approach. You should be confident of your abilities, able to assert yourself and not feel defensive when conducting a dialogue.

If you are aware of your weak points, you will know how to avoid falling into traps, and be better able to protect yourself.

For example, do you believe that other people are mostly straightforward and honest when dealing with you? Facing the truth, even when it isn't entirely pleasant to hear, will show that you are mature and able to cope with self-realization. If you want people to be honest with you, then the least you can do is deal honestly with them in return.

Other people's perception of your character is more often than not far less harsh than your own. But you can find this out only by asking someone for some honest feedback. How easily are you able to laugh at yourself? Psychologists say that being able to make fun of yourself means you are secure. If you can't laugh at yourself, you may have problems in dealing with others who could tease you in light-hearted fashion. Making a joke that shows that you don't take yourself (or your life) too seriously will often help to diffuse a tense situation. Be aware of when other colleagues use humour to good effect. It is best learned from experts because, used correctly, it can work wonders, while, clumsily employed, it can be detrimental.

4.6.2 Influences

What about the people who impressed you in your life? Everybody has influencers, mentors and champions. They could be personal contacts, or people who have inspired you despite the fact that there is no personal connection. People you read about in the press, those you associate with, friends or relatives who have known you a long time. If you have time to write down a list of the people who've had an effect on your life, do so. Go one step further and describe what they did and why their influence was so powerful. This is an excellent lesson in self-awareness.

4.6.3 Control mechanisms

How much control over your emotions do you need to exercise? Do you think you actually control your emotions well? Do you ever think about it? If you don't know how well you cope in stressful situations, try to recall how you handled a recent tricky issue. How would you behave if your new assistant managed to delete some vital information from your computer? Did you know it's neither healthy nor realistic to control yourself too much? Experts say it can be counterproductive. For instance, if you know a particular situation annoys you, you'll probably go out of your way to avoid it. This isn't always a good idea: evading uncomfortable circumstances prevents you from getting any practice in dealing with it. How can you become the sort of person everyone is happy to associate with if you never test yourself outside your normal territory?

4.6.4 The perfect solution

The ideal person, in general, is someone who manages to get on with all different types of people. They don't complain, don't show any behavioural excesses and seem to find everyone easy to deal with. Are these people just lucky in steering clear of trouble or are they experts in human relationships? People admire and respect them and, should any uncomfortable circumstances occur, these individuals seem instinctively to know exactly how to resolve them.

If you want to become a member of this enviable group, you need to have deep stores of self-confidence. Not for you the loud voice, trumpet-like, making the whole world aware of your talents and achievements. There is no need to go out of your way to impress others, you do so by your quiet efficiency.

What about when faced with a colleague asking you a favour? Do you squirm and wriggle like a fish on a hook, with no idea of how to refuse? If there are people in your office who assume the role of doormat, they are frequently people with low self-esteem, who have particular difficulty saying no. The main reason people find it hard to say no is fear of rejection. If you can learn to do it tactfully and gently, you'll be respected and gain some ground.

The more assertive types decline doing something with great ease. They do it quite politely but firmly and no one ever seems to mind. It doesn't mean that they never get asked to events, or invited to attend meetings, or give presentations. The irony is that, the more often you say no to unnecessary and irritating requests, the easier it becomes. The bonus is that it happens less frequently, and the requests you do receive are the ones you don't mind accepting.

4.7 Office politics and workplace relationships

People mean politics. Anyone who works in organizations recognizes this and should act accordingly. No one can ignore this reality of life and survive for long, let alone prosper. Office politics have predominantly negative connotations – the phrase summons up images of backstabbing, Machiavellian plotting and watching over your (or someone else's) shoulder.

To succeed in overcoming problems is a challenge. It's when circumstances and people combined conspire against you that the going gets really tough. There's nothing wrong with healthy competition between colleagues. Most people would claim that it's a good thing and a simple fact of human nature. A bit of friction isn't necessarily bad, either. It can act as a catalyst and stimulate creativity and output.

> **Networking hint**
> If you can make the necessary adjustments to minimize the negative effects that office behaviour can have, you'll find ways of harnessing behavioural differences of colleagues and make these work for you.

The worst aspects of working relationships are unconstructive. How many times have you found other people's disposition and behaviour affecting you? Isn't it hard work keeping up a sunny frame of mind while sitting next to a grumpy, scowling colleague? The ability to disregard other people's sighs, moans and groans shows that you are grounded and steady in your own personality. The last thing anyone needs is to have their effectiveness undermined by the selfish behaviour of a co-worker.

Developing a thick skin helps, as does distancing yourself from colleagues who are boring, unpleasant or moody. Don't allow yourself to become a prey to these formidable people. If they detect a chink in your armour, they will home in on it like a missile and exploit your weaknesses. This makes upsetting you a whole lot easier for them.

Accepting yourself as you are – most likely with no more than a few annoying traits – is what's needed to help you end up a winner when you come across irksome people. This is a sound basis on which to build healthy workplace relationships. You can always ignore other people's bad behaviour when you come into contact with it. Showing offence, being affronted, puts you in victim mode straightaway. This is an inferior position and one you don't want to assume. Avoid being manipulated by them, otherwise you can't take control, let alone gain the upper hand in the exchange. You don't stand a chance of coming out on top if you show signs of inferiority at the start. Keep cool and maintain your distance.

Before we move on to things in greater depth, there are some important factors to consider.

4.7.1 Intentions

People's behaviour always has a positive intention, whatever way it comes across to others. The problem is that their intention may be positive for them, but not necessarily for their colleagues. Everyone has their own agenda. Some of these can include interaction and positioning with others. Personal ambitions can be strong – here are some aims:

- getting the job done;
- increasing personal job satisfaction;
- organizing greater visibility for themselves;
- impressing others;
- securing greater rewards;
- gaining power;
- taking on more responsibility;
- beating others in a race for promotion.

This list shows the areas of potential conflict since a number of these points clearly involve competition. If one person gets more responsibility and greater authority and another does not, it will cause friction. These issues are all highly personal. To one person, being given more work involving, say, travelling abroad might be regarded as a perk. To another it would seem an ordeal, even though accepting the added responsibility would give either person an advantage.

Because of their personal agendas, people work to achieve what they want. This involves their pushing harder and harder to accomplish their aims. They may even go as far as preventing or handicapping someone else in their desire to succeed. This is where assertiveness becomes aggression, and brute force can result. What they do may be secretively devised, subtly and even invisibly deployed. But it is still a ruthless action and has no other purpose than to obtain personal advantage at another individual's expense.

Additional factors may also be at work. Some people are inherently less concerned about others than they could be. This may be because of their own feelings of inadequacy or incompetence (real or imagined). Some may have what is frankly described as destructive streaks.

4.7.2 Appearances can be deceptive

Because of the way people approach their jobs, and the degree to which they behave politically in the workplace, personalities may be deceptive. Things are not always what they appear. What is said may contain hidden messages and agendas. There are the usual known phrases to watch out for, such as 'trust me', 'let me be honest with you', 'I'm on your side, here'. The wary should try to read carefully between the lines as the communication continues. If you can determine people's motives, you will begin to interpret what they are really saying. Don't automatically think the best of them, or give them the benefit of the doubt. Prudence is a good watchword in most corporate environments.

It is useful to behave with caution. Keep an eye on the possible political implications of others' actions. Watch for dangers and opportunities because what you observe may provide an indication of either. Colleagues need to be assessed as potential friends, or enemies. Sometimes, confusingly, they may be one thing on one occasion and another at a later time.

4.7.3 Keeping a watchful eye

You should watch for, and read, any signs that could prove useful indicators. For example, notice:

- what is said;
- how it is said;
- alliances and changes of allegiance;
- people's intentions and motivations;
- the behaviour of others.

What you should try to do is develop a way of working with your colleagues that embraces the political environment. No one can work in isolation, so others can and will affect your progress. The question to consider is: are they assisting it or hindering it? What you need to work out is the action required on your part to ensure a positive outcome.

In all sorts of ways, this view will colour your judgement. Be careful when deciding whom you take into your confidence. How much can you confide in them and how soon? When should you ask for advice? How much should you publicise your success (or, conversely, hide anything that is less than successful)? Whom should you know and who should know you? What do you want other people to think of you?

All these questions are important. Each one requires that you make considered judgements and, above all, keep your ear to the ground and know what is going on. If you adopt the role of an out-and-out politician when building workplace relationships, this approach may do you more harm than good. Most organizations contain a few individuals who are recognized for their ruthless opportunism and politicking. It is sensible to avoid them, or at least not to cross swords or fall foul of them. Their ploys ultimately may do them no good if they are anticipated early enough.

4.7.4 Best foot forward

You have to decide how you are going to play things. Adopt a brash approach and you may gain nothing. Avoid or ignore all the intrigue and you may be left languishing on the sidelines, while others get the better of you. So what is the best way forward?

The plan you adopt will depend to some extent on the style of those around you, and the culture of the organization you work in. Openness, for example, could be an advantage in a normally secretive environment. However, it might upset someone who was incapable of anything other than a secretive approach. The way forward needs to include others, not only the prime players. Don't exclude those who remain neutral in the overall game plan. They could potentially help in specific ways. People add to your knowledge and competence. You could find a network of advisers, collaborators extremely useful.

> **Networking hint**
> Take people out of an organization and nothing of significance would be left. They form the essential ingredient of any corporate survival strategy. Having the right network cannot be left to chance. Your intentions should be decided early on.

You should systematically work towards making and maintaining the necessary contacts to foster the whole process. If this sounds scheming, that's because it is – but it is necessary. Without such a plan you will be defenceless against others. This is all part of organizational life.

People, contacts, relationships, interactions are what make working in an organization possible, interesting and fun. But the politics that go hand in hand with organizational working are full of potential risks. Worrying about who is on your side, who does not like you, who is seeking to score points and make life uncomfortable for you – including usurping your position – all provide reasons why you need to understand the way it works.

By taking a realistic (not pessimistic) viewpoint about behaviour at work, and a proactive approach to understanding how to deal with it, you can do more than just survive in the workplace. You can harness a number of positive and influential relationships that will keep you on track while you're at work. There will no doubt be a number of uncomfortable moments to be endured during your career. Sometimes you will not actually realize what you are getting involved with, or why you are doing something.

Your goal is to find a way of working that enables you to progress despite the deviousness of colleagues. If you do not fear confrontation and refuse to permit certain individuals to intimidate you, you will succeed. You will be able to operate effectively and achieve your own personal agenda. The right frame of mind is essential, and you should always focus on positive results.

4.8 Meetings and how to conduct them

Networking and meeting other people inside and outside the work environment are important to your success. Meetings are the most common form of professional communication and, should they be handled properly, there is nothing more effective. The problem is, despite the frequency of meetings you attend as part of your professional career, they are not everyone's favourite activity.

Most busy professionals complain about the number of times they are expected to attend meetings. Strangely, though, should someone learn that a meeting has been called and taken place and they aren't on the circulation list, they suffer pangs of mortification and anxiety. Why? Why were they not invited to attend? What exchanges of information have they missed? Was it because of something they'd done or, even worse, not done? Meetings themselves can be the cause of ambivalent feelings. People say they hate them in one breath, yet the moment they find they've been left out of a meeting group they immediately wish they'd been present.

Meetings can be extremely hard work, often boring, and can end up serving no useful purpose. Whether informal or formal, meetings should always serve a purpose – to advance a topic, disseminate information or prompt a decision. There is a saying: 'The ideal meeting is between two people, one of whom is absent.' So much of your time may be spent in meetings – this means there is a high cost element. No one will argue that they are an unavoidable part of organizational communication, consultation, debate and decision making. They are necessary – but *all* of them?

4.8.1 Meeting fatigue

Are you spending so much time in meetings that you have no time actually to do any work? Do you measure your day by two-hour appointments that invariably run into each other? Is your briefcase packed full of minutes to be read and action lists to distribute? If the answer to these questions is yes, then you are suffering from meeting fatigue. This is a widespread phenomenon in many industries and professions. It can lead to all sorts of frustrations. There's 'voicemail rage', where the caller is so incensed at not being able to speak to a real person that the machine takes a battering of abusive language; there's 'minute taker's syndrome', the telltale laptop-bag slouch and ability to type and talk at the same time; there's 'speed-read disease', where minutes are read so swiftly that key actions are missed with occasional disastrous results; not to mention the caffeine dependency introduced by non-stop cups of coffee.

4.8.2 A communication tool

Meetings are a vital part of how networking and relationship building can be put into practice. But they should be an aid to help you. Individuals need to be able to talk things over with colleagues, to bounce ideas around within their own organizations, to access databases of information, to contact suppliers and specialists for advice. All these things require each team member to have their own time and space in which to pursue their discipline.

So what do all these meetings actually achieve? Client meetings should be about reporting on the progress of the job, allowing the client to ask questions. A well-executed client meeting should be boring, as there should be no new information, and it should be swift. There certainly are clients who will want to examine issues in detail, but they should be encouraged to do this at a more appropriate forum.

Dissemination following meetings often leaves a lot to be desired, too. If meetings are sandwiched together, then reporting back gets squeezed out, and others in the project team are left to carry on in the wrong direction, perhaps unaware of vital information. One suggestion is to send apologies to every other invitation to attend a meeting – you'll spend more time in the actual process of getting something done and less time doodling on the back of meaningless agendas.

4.8.3 Effective meetings

Because you could spend a fair proportion of your time in meetings and it is essential to behave appropriately if you are putting your relationship-building skills into practice, remember that effective meetings do not just happen. Everyone's role is important, whether they are running the meeting or attending it. If you are 'in the chair' here are some basic rules to follow.

Meetings as a form of communication can be used to:

- inform;
- analyse and solve problems;
- discuss and exchange views;
- inspire and motivate;
- counsel and reconcile conflict;
- obtain opinion and feedback;
- persuade;
- train and develop;
- reinforce the status quo;
- instigate change in knowledge, skills or attitudes.

The key role is surely to prompt change. There's not much point in having a meeting if everything stays the same. Decisions need to be taken, and this means that a meeting must be constructive. If you can develop a reputation for organizing a good meeting, people actually will want to attend it. They will be keen to do so because it:

- keeps them informed and up to date;
- provides a chance to be heard;
- creates involvement with others;
- can be a useful social gathering to allow cross-functional contact;
- provides personal visibility and public-relations opportunities;
- can broaden experience and prompt learning.

Meetings are potentially useful. The progress of a project can, in a sense, be certain only if meetings are held regularly and are productive. Making meetings work – like most things – requires planning. First ask yourself these questions:

- Is the meeting really necessary?
- Should it be one of a regular series?
- Who should attend (and who should not)?

Once you have satisfied yourself on these points, you can proceed to the following.

- Setting the agenda: This is very important. No meeting will go well if you simply make it up as you go along. Notify people of the agenda in advance and give good notice of contributions required from others.
- Timing: Set a start time and a finish time. Then you can judge the way it is conducted alongside the duration and even put some rough timing to individual items to be dealt with. Respect this time schedule.
- Objective: Always have an objective so that you can answer the query, 'Why this meeting is being held?' The answer is not, 'Because it's been a month since the last one.'
- Preparation: Read all the necessary papers, check all details and think about how you will handle both your own contribution and the stimulation and control of others.
- Encouraging others to prepare: This may mean instilling habits in attendees (punctuality, concise contributions from among attendees and not pausing for someone to read documentation that should have been studied beforehand).
- People: Who should be there, and who should not? What roles do these individuals have?
- Environment: A meeting will go more smoothly if people attending are comfortable and there are no interruptions.

Leading a meeting does not have to be done by the most senior person present. The chair does not do most of the talking – in fact the reverse is true. The chair should be responsible for directing the meeting. Effectively conducting a meeting ensures that:

- the meeting focuses on its objectives;
- any discussion is constructive;
- a thorough review can be relied on before decisions are taken;
- all sides of an argument are aired, reflected and balanced;
- proceedings can be kept professional and non-contentious.

A good chair will see that the meeting progresses correctly, handling the discussion and acting to see that objectives are met. It is essential that only one person speak at a time and that the chair decide who (should this be necessary).

It may be necessary to prompt discussion of certain matters on the agenda. Let us look at some ways of doing this.

- *Overhead questions:* These are put to the meeting as a whole and whoever picks them up starts the discussion process.
- *Questions direct to an individual:* Without preliminaries, this is to get an individual reaction or check understanding.
- *Rhetorical questions:* A question demanding no answer can still be a good way to make a point or prompt thinking. The chair could provide a response if desirable.
- *Redirected questions:* These are questions asked of the chair but batted straight back to the meeting, e.g. 'Good question – what do you all think?'
- *Developmental questions:* This is a prompt to get discussion started and builds on an answer to an earlier posed question and moves it on. 'Right, Jim thinks this is too expensive. Does everyone agree, or are there other options?'

Prompting discussion among the other professionals present is as important as control. It is the only way of making sure the meeting is well balanced and takes in all required points of view. It prevents someone returning at a later date with a remark along the lines of, 'This is unacceptable. My company wasn't given the chance to make any representations earlier.'

Networking hint
Keeping control is essential. Do not, if you are chairing the meeting, get upset or emotional. Try to isolate feelings from the issue itself. Agree that it is a difficult point and concede that feelings will run high.

Maintaining direction is imperative – if matters are escalating, call for silence or a short break before moving on. You could suggest putting the problem to one side, or asking for a subgroup to be formed to deal with the tricky issue on its own. As a last resort, abandon the meeting until another time.

At best, meetings should be creative, fostering open-mindedness between the involved parties. The person in the chair should actively stimulate creative thinking by ruling against the instant rejection of ideas without consideration; should contribute new ideas themselves or steer the discussion into new directions; should find new ways of looking at things; should consider novel approaches and give them a chance; should aim to solve problems, not tread familiar pathways.

Some groups who meet regularly manage to do this spontaneously, but others do not. For those who fall into the latter category here are some prompts.

- Make it a rule that creativity will become part of the culture of the group or project team.
- Make it easy by providing feedback from the group to build in an ongoing exchange of ideas.
- React to it by acknowledging what people say. If it is useful, thank them; if it is negative, tell them why and suggest other approaches.
- Give credit openly, in public, because nothing will ensure a flow of ideas more than praise; and make sure it is channelled upwards as well as laterally.

- Make time to deal with creative ideas so that it is seen that consideration has been given to these points.

Finally, if it helps, arrange meetings as part of your networking culture. They could take the form of 'discussion lunches' at midday, with sandwiches. This might encourage higher attendance with a good mix of people, and facilitate idea generation, problem solution and identification of opportunities.

Chapter 5
Rising to new challenges

Are you worrying that building successful relationships at work is just a bit more than you bargained for? If you are, it shows that you are taking things seriously and have ambition. No one ever got anything without effort. Your new challenge could take you a step (or two) up the career ladder. However daunting it seems at present, all that is required is focus, application, specific approaches and new skills. Things can be taken at a steady pace, and adjustments made where necessary. This book is designed to help you feel able to deal with each step as you go forward.

It is advisable not to underestimate the amount of change involved in the transition from being heavily task-aware to being more people-oriented. Even if it is not an entirely new concept for you, your decision to work as hard at getting on with people as you have been at dealing with tasks will almost certainly involve added responsibilities and challenges.

Here are a few points to bear in mind. Do you work in the private or public sector? The cultures within these different organizations are marked. Depending on where you are employed, the networking culture should be germane to the organization. It is advisable to pay particular attention to what is involved in your specific situation.

Perhaps you work for a global company. This could involve cross-cultural working relationships. As the UK is one of the most multicultural nations in the world, and London has one of the greatest mixed populations, you are likely to be working with a number of people from across a variety of continents. Where workplace relationships are concerned, it is a matter of different strokes for different folks. You (and your colleagues) may need time to adjust to this, particularly if you are in a new job where things are very different from anything you've experienced before. Many organizations offer excellent induction training and you should take advantage of whatever there is.

Depending on your level of seniority, your training manager, or HR director, is the person to turn to for help and advice. Some organizations offer formal 'induction' programmes. You may have a number of meet-the-people series of events arranged. The induction timetable is adhered to until completed, by which time total absorption in the corporate culture and hierarchy should have happened.

It is possible (though hopefully unlikely) that you could be thrown in at the deep end and you could be struggling quite quickly. If you are not given any induction training, request some from your immediate superior. It is only fair that you be given adequate time to adjust to the new working culture and hierarchy, so that you can begin to work out what rapport building will be needed to feel comfortable in your job.

Continuing with the new-job theory for a moment, people achieve a new position because they are good at something. How did you get your current job? Did you get it because you were particularly successful at what you used to do? Have you moved to a more senior position within the company, or to a new one, doing broadly similar work because you have outstanding experience? Or did you get your present position because of the skills you have?

If you want to be successful at building workplace relationships while being effective at what you do, the most important issues you will have to face are the following.

- Challenges: Do you have enough/too many?
- Learning curves: Do you welcome them/do you learn easily?
- Personalities: How do you cope when there are clashes?

One of the best ways to ensure you succeed in your goals is to have the right **attitude**. If you see relationship building as something exciting and different, requiring new approaches and offering you greater challenges, this is a healthy sign. You may have some previous experience (good and bad), but by looking forward, trying to learn fast, you will be showing the right attitude.

> **Networking hint**
> One of the most important things in the beginning is to keep an open mind about your colleagues. Suspend reaction, try not to make assumptions and judgements, or jump to conclusions about systems they use, how they deal with other people or how they run projects.

You should consciously define and adopt new ideas and ways of approach. Your **actions and attitude** can make all the difference between success and failure.

Without doubt the main challenge you face is **other people**. You may prefer to work on your own, because, if you are particularly task-oriented, this may be how you got to where you are. But, if you are trying to build rapport with others, one way of doing this is working as part of a team. You could work in an open-plan office where people are interacting all the time. It is likely that, in the course of your work, you report directly to others or have staff reporting to you.

Working with and relating to people requires effort; it takes **time and expertise**. There is much more to it than simply getting on with a task. It is important to remember that tasks can be carried out, skills taught and new habits acquired relatively quickly. Results are what it is all about. But, where people are concerned, although you need to find out swiftly and sensitively about their attitudes and personalities, the process takes time. Getting along with colleagues is a long and often quite complex process.

Does your position involve a set of **key tasks**? These could be:

- **planning** (what must be done to achieve the desired outcome);
- **organizing** (yourself, time, other people and activities);

Rising to new challenges 71

- **recruitment and selection** (if managing a team or department);
- **training and development** (your own skills or those of your staff);
- **motivation** (creating and maintaining a positive attitude);
- **control** (monitoring your own performance standards and those of others).

Success in these will vary depending on the exact nature of the work you do and how good you are at dealing with other people. It is most likely that you will be expected to be able to take decisions and make them work, as well as being adept at problem solving, time management, communication skills and relationship building. It is essential to work hard at becoming expert in the skills your role demands.

5.1 Three aids to a networking strategy

Three things will help you when you start your networking/relationship-building strategy.

5.1.1 Aid 1: Confidence

It is easy to feel intimidated and overwhelmed when starting something new. However, if you approach people and relationship building with confidence, you will find that some colleagues will trust you. Courses in such subjects as public speaking and presentation skills can help you overcome shyness and build your confidence in the early stages.

5.1.2 Aid 2: Knowledge

Learn people's names. If you've been seriously task-aware (head in the sand) for a long time, you could be at a disadvantage here. Make a point of finding out who your company's directors and other senior personnel are and their roles. If you're new to your job, or simply haven't had the time, check the company website often for updates. Most now include pictures, so you can see what people look like. If your organization has an intranet, and most do, use it. It can be a great resource. Everything you need will probably be found on the intranet site.

To be good at interpersonal relationships at work, get as much information as possible at your fingertips: internal phone directory; speed-dialling codes; departmental electronic diary; contact lists; useful phone numbers; operation manuals (voicemail, telephones, etc).

Now you've made a start, socialise as much as possible. Accept invitations to after-work drinks, lunches and group events. Arrange coffees and lunches to get to know your immediate colleagues better. If time allows, volunteer to help with work events/corporate functions. The sooner you get involved the better.

The greater the interpersonal skills you have directly affects your ability to succeed in your role. Some people say that success in business is 20 per cent strategy and 80 per cent people. If you can communicate well with others (business writing, making effective presentations, running meetings, liaising with staff on a one-to-one basis, conducting interviews and appraisals) use these skills wherever you can. This will improve your effectiveness sooner rather than later.

5.1.3 Aid 3: Focus on the positive

However difficult you are finding this (and however reluctant you are to practise), make a note of the skills you have and how you rate them. This sort of self-analysis is useful when you are trying to measure your progress in building rapport in the workplace. You may have received feedback on previous occasions and found it accentuated the negative. Employee evaluations tend to focus on opportunities for improvements, with the sting of criticism often lasting longer than the faint glow of praise.

Maybe it would be sensible at this stage to conduct a self-appraisal designed to give you an idea of your progress in developing your people skills. Perhaps you have a couple of people who could help you do this. Collect some feedback from people who know you well – inside or out of work. But ask only for positive information on your people skills. When you've spoken to them, what are the results? Do they say broadly the same things?

Spend a bit of time working out the areas you think you could improve on. The purpose is to build on what you are already good at and find ways to work on the other areas. By using your strengths you can shape the pattern of progress towards the next phase. Taking action as a result of accurate self-appraisal is a very honest way to help yourself achieve your goal.

5.2 Focus on priorities

Whatever you are trying to do, you can always apply Pareto's Law (the 80/20 rule). This states that 80 per cent of results flow from just 20 per cent of causes. Where relationship-building skills are concerned, this means that you have to get only 20 per cent of what you do right to achieve 80 per cent success. Now that doesn't sound too bad, does it? Your position as a keen networker will be far easier if you concentrate on the right things first off. This will have great impact on future results.

So, as early as possible after starting your networking strategy, concentrate on your own priorities. Make it a rule to manage yourself, and work in a way so that you focus on your core 20 per cent of key tasks. Your objective is to have a **productive and effective** network of contacts within the company. If you are required to manage others, try to get them to work in a way that reflects the realities of the 80/20 rule.

> **Networking hint**
> Your networking success depends not only on yourself but also on the other people with whom you work. Other people will relate to you better if you are good at what you do.

Going back to the importance of creating a good first impression, do you appraise other people when you first meet them? It does not take long to form an initial view of someone. Most people will:

- **observe** the person's manner and style;
- **listen** to what they say, then read between the lines;
- **watch** what they do and how they do it;
- **look** at how it affects and relates to them.

It's quite unnerving if you think about this too much. If you are in the position where people are likely to observe your every move, it is important to *prepare as well as you possibly can* before you start and then *start as you mean to go on*. If you're intent on making a good impression on those you work with, *you* have to make it happen. No one else can do it for you.

5.3 A good beginning

Success is something that is actively gained, not achieved through good luck. There is a saying: 'Luck is a matter of preparation meeting opportunity.' It is always a good idea to take advantage of any lucky opportunities that come along, but, without a success strategy, luck alone will not get you where you want to be. Once you've devised your internal networking strategy you are aiming to demonstrate to superiors, colleagues and staff that you are well positioned to succeed.

Your action plan could take two forms: self and task. Setting up an action plan will help you to remember what needs to be done, how it should be done, when it needs to be done by, and the impact it will have on whom and on what. In its simplest form, your self-action plan could be as follows:

- **skills** (yours and those of the people with whom you are building relationships);
- **personality factors** (where there are similarities and contrasts);
- **knowledge** (of their job, role, organization, people, product – whatever is relevant);
- **connections** (whom *they* know may be as useful as whom *you* know);
- **profile** (how they are perceived in the organization);
- **attitudes** (how they affect their work and dealings with other people).

If you match this list with your own personal attributes and those of your new contacts, you will see how things begin to shape up. How do you compare when lined up alongside these people? Are you at all similar, or are you total opposites? Where are the complementary areas? Do you have skills that need strengthening in order to build a relationship with them? Which aspects of your personality give you an advantage? Which ones will you have to curb? Which areas of knowledge do you need to extend? How will you do this? Whom do you know who might be helpful to you in this relationship-building strategy? How easily will you be able to establish rapport with them?

It might be sensible, when considering your answers, to take a short-term and then a long-term view. For example, you may well have some skills that will enable you to make some 'quick wins' in your networking strategy. However, there are some people whom it will take time to get to know, so patience will be necessary.

Your task-related action plan might be to:

- **read** everything you can about the people with whom you want to network;
- **attend** courses if there are any that relate to developing people skills;
- **persuade** someone (a member of staff, colleague) to mentor you;
- **obtain** permission (from your manager) to attend internal events.

If you can develop the habit of sketching out an action plan (self- or task-related) daily, weekly or monthly it will be a useful aid to people-skills management, self-organization and effectiveness.

5.4 More reconnaissance

You need to be aware of what information will be useful to you where relationship building is concerned. Also, you will have to work out where you can find it. Here are some things to consider:

- **targets** (self-imposed personal and group/departmental);
- **procedures and systems** (what should you be familiar with ahead of time);
- **people** (who's who; should you meet any of them before you start networking with them; do you know anyone who knows any of them?);
- **lines of communication and reporting** (whom you report to and who reports to you, what processes are used – meetings, written reports);
- **controls** (how will you monitor your performance and progress?).

Make notes of everything in your database, as you acquire more information. The more confident you become, the easier this will be. You will come across as a responsible and reasonable person. What you are trying to achieve here is the best possible position from which to launch your networking strategy, so that any future dealings will not be too problematic.

5.5 Taking the plunge

You may have taken only a few tentative steps at networking up to now, but let's recap on the reasons why it is necessary to have people skills. What use are they? Why should professionals in particular need these skills and how do they go about making connections?

For a start, there are countless opportunities to meet fellow professionals when networking externally. Socialising is something that most of them enjoy. Should you be a bit shy or something of an introvert, you are bound to be a little fearful of what you're getting into. Networking may (in the abstract) not seem the sort of thing you feel confident about. Are you convinced that luck will be against you, that you'll meet only unreasonable, unpleasant or dull people?

If you are apprehensive about the networking process you'll find it more than easy to justify not doing it. Think of the advice given about procrastination – it is equally applicable here, too. You will be able to think of lots of seemingly valid reasons for avoiding social contact, but they're not reasons really: they are actually excuses. You can't get there on time; that day you're really busy; surely you'll be able to find someone else who can go; perhaps you'll do it next time – in a month, or next year. Many people faced with doing something they dislike swiftly find evasion techniques they never knew they possessed. It's best, if you can, to take stock of the situation honestly and do an appraisal.

The safest way to start is by building relationships at work within the organization. Then, when you are sure in your own mind about where you want to go and how you make your initial approaches, you can do so. How can anyone but you know whether you're comfortable with a situation or not? In particular, where you are trying to facilitate new contacts in your profession, it will be a bit of trial and error in the early stages. No one else can run your networking campaign better than you. So start off inside the workplace until you gain confidence.

Networking hint
Wherever you begin your forays into the networking arena, your first success will probably be in the most unexpected place at the least anticipated time. You may feel most relaxed pursuing casual methods of meeting people, rather than just attending company events and meetings.

Many organizations have a policy whereby they get involved in philanthropy or good causes. It may be that your company sponsors a charity or a fundraising event. You could quite easily volunteer your time and energy (if you don't have much money) as a charity volunteer, or by participating in fundraising events. If they have the blessing of your boss, you will find this an excellent way of making new acquaintances as well as feeling good about making a contribution to a worthwhile charity project. The more involved you become in whatever voluntary work your company encourages its employees to do, the better you'll get to know others who share your interest and desire to 'make a difference'. This is where the personal touch comes in. You will be far more enthusiastic about working as a volunteer if you have passion for the cause.

Many wonderful organizations offer volunteer opportunities but some of these require a serious time commitment. Take into consideration that you will not be paid, and, after all, it is your professional life that comes first. Perhaps that is why the most obvious place for networking opportunities is through work. After all, where do you spend the majority of your time each week? Sometimes, because you see the same people every day, familiarity blinds you to the fact that they could be useful people to include in your network of contacts. If you are thinking of networking to enhance your career, the workplace is the most sensible place to start. You could meet a whole new crowd of people if for example you've recently moved to a new office location, or you've just been promoted to another department. Some companies have great opportunities for networking, so don't overlook anything.

5.6 Mix 'n' match

If you have no fixed preference for a particular method of starting your networking campaign, it is quite OK to do whatever you're comfortable with. Remember that people tend to be judgemental; you make up your mind – usually within a few seconds – as to whether or not you wish to pursue a conversation with someone. With networking, it's not a good idea to be too hasty or dismissive. Other people may be shy and not appear to advantage at the first encounter. It might be safer to keep your options open and give things time to develop. There is no limit to the number of approaches to networking you can adopt. It's your choice entirely, limited, perhaps, in the workplace only by time. The most important thing is to do only what you feel comfortable with and remain in control at all times.

Given the many different types of people there are all over the world, you will have definite ideas as to what sort of people you like. Try to mix the methods you use, because this will optimise your chances of success, you will build confidence faster and you will be able to judge which feels right for you. Once you start on this process, it will become easier for you to identify people with whom you have common interests, and with whom friendships may develop. Whatever you do, make sure you feel comfortable doing it.

5.7 How to win friends and influence people

If you haven't read the famous book *How to Win Friends and Influence People*, written by Dale Carnegie and first published in 1936, you may like to know that it's still available and well worth browsing through. Certainly, for those keen to add networking to their list of skills, there is nothing more important than being able to make friends with your work colleagues. A friendly wave or smile as you enter your place of work in the morning can transform it for you in an instant.

Whom do you know in your company? Where do they fit in the office hierarchy? If you can already identify them and have an idea of their job titles and what they do, you won't waste valuable time and resources dealing with them in the wrong way. Despite having been, up to now, rather task-aware, do you know who among your immediate colleagues and department will be pleased that you are emerging from your self-imposed hermit-like existence? They are definitely the ones to start with (while you are feeling a bit apprehensive), to engage in casual conversation with or ask if you might join them in their lunch break.

Hopefully, you know who the real decision makers are in the company (not those on the organizational chart but those who wield major influence in the boardroom). What is the basis of their power? Is it time spent in the company or previous relationships with one or more members of the board, or are they major shareholders?

Remember, everyone in the company is important – don't underestimate anyone just because they are junior as their power base may be much greater than you think.

No doubt you have, as a keen networker, already started to fill in your contacts database, which includes your colleagues, staff, team members and superiors. You could usefully spend a bit of time working out what sort of people they are. If you can develop the habit of doing this when you meet people, it is amazing how easily and successfully you can connect with them on something you share an interest in. Once you've identified a particular type, you will find that you meet other people who resemble them. Whether they are physically similar, temperamentally alike or just have the same attributes, you will probably know instinctively how they'll react, what they will be like when you speak to them and how you will get on with them.

Case study

In his new position as finance director, Les found a woman had a direct route to the board and always knew what was going on, well in advance of her more senior colleagues. How? She had worked for the company for many years and had comforted (not in the physical sense, but had been a listening ear to) the CEO when his wife (whom she knew) had died several years previously. So, unknown to the staff, there was a special working relationship between her and the CEO, who made sure she was in the know on business matters. Les says it took him some time to find out that she had knowledge that might be considered above her pay grade, and he nearly made some costly mistakes in the meantime. In the end, she and Les became great allies, but things *could* have gone catastrophically wrong.

Companies prosper when staff are genuinely interested in their colleagues and others. When trying to build relationships with people, it is important to remember that the process requires confidence. Some people are naturally reticent while others are born extroverts. People who are skilled at this show certain characteristics.

- They treat everyone as being interesting, special and likeable.
- They use good eye contact and positive body language.
- They make other people, particularly new acquaintances, feel special.
- They introduce people to each other effortlessly, remembering names and something relevant about those they introduce.

In other words, they have 'charisma'. Charisma counts: if you are charming, it is contagious – you come across as being generous and will be able to build self-esteem in others.

You probably recall having met someone like this already in your career. When you were introduced, they smiled, entered into conversation easily and drew you out. They probably asked you questions about yourself, and listened to what you said. In essence, they made you feel important. When you parted, you probably thought to yourself, 'What a great person!' Not only is it easy to be in their company, but these people are at ease with themselves.

If you work with someone who is highly task-aware (or even if you are a bit that way inclined yourself), you will find that such people often find it difficult to appreciate the value of personal relationships at all. They are happy to sit at their desk, working at their computers all day. This is often preferable to interacting with other people. As a colleague, you are likely to receive an email request from them, rather than a personal approach, even if they are sitting next to you. They may go so far as to avoid the coffee machine in case they get caught up in conversation. Sometimes they cover up their reticence to interact by saying they have too much work to do. Don't underestimate the importance of a bit of staff bonding. It goes a long way to encourage good relationships among team members and colleagues. If you can cultivate the ability to see the world from other people's perspectives you can then work out how your new colleagues prefer to work.

Do you know how much staff members really know about each other? Can you recognize who the key people are within your organization to network with? It sometimes takes a while to work out who the decision makers, the movers and shakers, and the influential persuaders are.

But, in addition to these well-connected people, it is important to work out where everyone fits in to the hierarchy.

You may not have thought about all this, because you're too concerned about how to begin building good rapport with people. Don't forget that your colleagues and your superiors are human beings as well. They have hopes, fears and insecurities – and sometimes they need nurturing too.

5.8 Identifying key players

5.8.1 Influential people

It is a common assumption – and you'd probably be surprised at the number of people who believe this – that influential people are those who hold high office. This isn't necessarily so. If you have a wide range of contacts, some of whom go back a number of years, you will find that some are not in elevated positions, or immensely senior, yet they wield considerable power.

> **Case study: Going under**
>
> The following is extracted from 'Change Management', in my book *Making Management Simple*, published by How To Books of Oxford.
>> The board of a company decided as part of their modernization that they needed to join two buildings together which were separated by a busy road. They commissioned

Case study: Going under (cont.)

architects and consultants to apply to the local planning department to build a bridge between the two factories. The application was refused. They spent many hours and much money researching other solutions but came up with none. The board and the consultants were stuck.

One morning the chairman of the board was driving to work, and saw ahead of him the caretaker on the other side of the road. The man disappeared into the building but by the time the chairman passed that spot he saw to his amazement the caretaker standing on the other side of the road. The chairman stopped his car and shouted to the caretaker, 'How did you do that? Get from one side of the road to the other without walking across?'

Answer? There was an underground maintenance passage. For some reason it was not on the building sites' plans but it was in daily use by a small section of the workforce.

Moral: never be too proud to ask or underestimate the knowledge and experience of every single person you work with.

Here is another example (though this one is not from *Making Management Simple*).

Andrew used to work for a global company which was hierarchical in its approach. One day he needed the advice of the Chairman on a particular matter. He went through the usual channels and asked the Chairman's PA if he could possibly see him for ten minutes. She replied that he was not available for two weeks.

Andrew knew the decision could not wait, so he went to find the Chairman's chauffeur. He found him in the company canteen and bought him a cup of coffee. In the course of the conversation he asked Charles, the chauffeur, where the Chairman was that day. Charles told him he was driving him to the airport that afternoon at 3.00pm, and if he wanted to see him he should be waiting by the main entrance. Andrew was at the door, as suggested, at the appointed time. The Chairman spoke to him, invited him to ride with him to the airport so that he could give the matter his consideration. The advice was given and the problem solved.

Some points to remember:

- be prepared to think laterally to solve problems;
- remember that a small piece of information can make a huge amount of difference;
- powerful connections don't mean just getting to know 'the great and the good';
- pay attention to everyone – and discover their individual strengths;
- sometimes a valuable piece of information can come from the most unlikely source.

5.8.2 Movers and shakers

These people usually far exceed the boundaries of their office positions. When beginning a networking strategy, it is wise to find out who they are in your organization.

You won't have trouble spotting them because they make it their business to know everyone and be seen. Movers and shakers are important to keep track of – you never know where they are going to turn up next.

5.8.3 Corporate citizens

These are the hardworking, non-political staff who are great resources for information and advice on almost everything to do with the organization. They have probably been around a while and know all the details of the company, their department and most personnel. They probably even know the date of the CEO's partner's birthday. Don't hesitate to seek their advice when appropriate. Better to ask them than find them saying afterwards, 'If only you'd asked me, I could have told you that . . .'

Sometimes it is easier if you can mentally sort people into categories. These are fairly sweeping generalizations but it may save you a bit of time if you're sifting through a lot of personnel.

'Road runners' are often highly task-aware and don't want anyone in their way of achieving targets. It is wise to let them go at their own speed because if you interact it will only slow them down.

'Race horses' get things done fast but like others alongside to help them. They are perfect to team up with because of the accelerated pace at which they work. If you join them you will find yourself flying along. A racehorse is a valuable asset in any group. They are strong and capable and can achieve great things but work best when there is someone in charge of the reins.

'New pups' are *the* most people-oriented types and will be so eager to please when you tell them that you are starting a rapport-building strategy. They like working with others but have a tendency to forget the importance of getting things done. They sometimes spend too much time being helpful and friendly.

'Tom cats' prefer to work on their own. They can be rather difficult to build rapport with. They are independent, and don't regard other people and teams as very important. Some of them produce the most amazing results but are often remote figures and content in their own company.

5.9 What colour are you?

People have different energy levels and there is a theory that they can be represented by a colour. What colour are you and your colleagues? There are four main types.

- **Cool blue**: These are usually regarded as stand-alone types. They can be cautious, precise, deliberate and formal. If they are a bit distant it's nothing personal. If you are trying to work with them, or influence them, you will need to work on them slowly.
- **Fiery red**: These are pretty much the opposite of the blues. They are competitive, demanding, determined and strong-willed. They will reach their goals whatever it takes. Sometimes they would do better to tone down their actions because they can overwhelm people.

- **Sunshine yellow**: Sociable, dynamic, demonstrative, enthusiastic and persuasive, the yellows are an asset to any organization, have natural charisma and are able to shine in any situation. They help keep morale high among colleagues; no matter how difficult their job is, they see things in a positive light.
- **Earth green**: These are the caring, sharing, encouraging, compassionate and patient individuals. If you've got any greens in your office, you'll find that they're the ones who always have the headache tablets and bandages, they water the plants and remember everyone's birthdays.

> **Networking hint**
> If you need to get on with people who are not like you, try adopting the chameleon approach: change colour to suit the environment you're in.

5.10 Watch the body language

If you have ever studied neurolinguistic programming (NLP), you will know that one of the beliefs is that you can never *not* communicate. People don't always realize that they give a lot of information about themselves without even opening their mouths. Body language can speak volumes. Equally, recognizing and understanding body language in others is important in relationship building. Paying attention to this may reveal what your colleagues think about you. Here's a brief summary.

5.10.1 Eyes right

Eye contact is important when dealing with other people. When meeting colleagues, maintaining eye contact shows respect and interest in what they have to say. It's appropriate to keep eye contact about 60–70 per cent of the time. This differs in other countries and cultures. If your expression remains riveted on them (because of nervousness) people assume there is something wrong.

5.10.2 Posture

Hopefully, your natural posture is good. Head upright and shoulders back shows that you are in control, professional and efficient. Slouching, drooping, shuffling – all of these inhibit breathing and emphasise the fact that you are nervous or uncomfortable.

5.10.3 Heads up

The position of your head is a great giveaway. When you want to give an air of confidence, keep your head level both horizontally and vertically. The straight-head position conveys authority and lets people know you want to be taken seriously.

If you tilt your head just a little to one side or the other during a conversation this indicates that you are interested in what the other person is saying.

5.10.4 Disarming yourself

Did you realize that arms give away clues as to how open and receptive you are? If your arms are relaxed and at the side of your body, you are open to whatever comes your way. Arms crossed tightly in front of you is a defence mechanism and indicates that you are protecting yourself from whatever is going on.

5.10.5 Legs

Because they are a long way from the brain, they need firm control. If you are nervous, don't allow your legs to let you down. Keep them as still as possible, particularly on first-impression occasions. When trying to build rapport you need to come across as composed and confident.

5.10.6 Hands

There are so many gestures, but remember that palms upward and outward are seen as open and friendly. Palm-down gestures are dominant and emphatic, particularly if the arm is held straight with no bending of wrist or forearm. When handshaking, make the gesture upright and vertical to convey equality.

5.10.7 Distance

This is crucial for people wishing to give off the right signals. Stand too close and you'll be considered pushy, too far away and you'll be judged as aloof or withdrawn. Neither is advisable, so best watch others to see how the group relate to each other. If you find a colleague backs away when you move closer to them, this is because they interpret you as encroaching.

5.10.8 Ears

Of course, the most important thing to remember is that you have two ears and one mouth, so use them in that proportion. If you keep this in mind, it will help communication with colleagues.

5.10.9 Mouth

This is an expressive part of the body. Lips can be pursed or twisted if you are thinking hard. Mouth clamped shut may give the impression that you are not pleased. Smiling is what the mouth is best for. Smiles create a feel-good factor where rapport building is concerned.

While you are trying to build relationships internally, it pays to spend time finding out people's likes and dislikes. A 'thank you' never comes amiss. If praise is due, then, if it can be done appropriately in public, say something – the results can be dynamic.

5.11 Finding the right opportunities

You are looking at ways of building rapport within your organization. There are some useful 'O' words to consider in this connection. Seek **opportunities** to make connections with colleagues on every **occasion** within your **organization**. What are the main **objectives** you have for building relationships internally? If you are intent on being popular you may need to make yourself **obvious**, but in the right way. As mentioned earlier, confidence counts. In order to make friends, **offer** people something. Always keep an eye out for **openings**, to get involved with new projects or incentives. An **ongoing** effort will be required but, as you gain in confidence, it will become easier to do. The more adept you become at making friends and winning people over, the easier you will find it when trying to **overcome obstacles** that may hold you up.

5.12 Some strategic advice

It is important to work out who's who within the company, which people you need to influence and whom you will ask for support when you need it. One simple and effective way to help you do this is to create a **stakeholder map**.

> **Networking hint**
> Creating a stakeholder map is a useful exercise if you work for a large organization and need to know the right people.

List on a piece of paper whom you need to get to know. Think about the ways you can currently make contact with them. Are there other ways of doing so? Have you some idea as to how often you should keep in touch with them? What opportunities exist currently? Do you need to create new ways to connect with them – either in or out of the office?

5.13 Creating a virtual team

Once you have decided to build your own personal network, you should try to identify your own 'hot list' of contacts. This is called your **virtual team** and consists of an 'inner circle' of colleagues who can help you and influence others at times when support is needed.

It is sometimes vital to survival, particularly in organizations where there can be fundamental changes in management structure. You could start to work on it as soon as you can. To begin with, you may well be relying on 'old connections' from a previous workplace, education establishment, family, friends, etc. They are usually people you know extremely well and have good advice when you come up against problems.

You will find there are probably no more than about six to ten people who can be included in such a close circle of allies. The importance of your virtual team is that they are your first port of call when you need advice. They will be interested in you and they understand the work you do. You will in time get to know their likes and dislikes and be genuinely interested in their success and happiness. Whatever it takes, you should keep in regular contact, even if they are transferred or promoted to another position or department. You will continue to be alert to opportunities for introducing them to new people. If there is a distance issue, keep in touch by whatever means are most effective.

Members of your virtual team should not be called on just to help with your problems. If you want to keep your visibility high, inform them about what you are doing, and ask about any new developments in their career as well. The unique thing about your virtual team is that you'd stick up for them, write a glowing reference, support them, listen sympathetically to their concerns, spring into action for them – whatever it takes. In essence, you trust them, because they would do the same for you.

Chapter 6
How to get (the communication) going

You're about to have an important first meeting with your team on a new project. It's an essential first step in the internal relationship-building process and your communications skills will be tested to the limit.

Face-to-face meetings can result in awkward pauses and initial shyness for those who are not brimming with confidence. To help you over this hurdle, you can approach the meeting fully prepared if you have a look at the following factors. In order to get your message across, think about what you are trying to achieve during the dialogue.

- What information needs to be conveyed?
- What do you want the others present to do as a result?

Organize yourself beforehand. Jot down notes about your major points. Be positive and keep the message simple. That's easy enough, isn't it? It's straightforward in theory, but in practice often fraught with difficulty. This is particularly so when you have high expectations from other colleagues present.

Communication is not just speaking, writing or gesticulating. It's more than the transmission of information. Something else has to occur for the communication to be complete. In essence, the other parties to the communication method have to engage the brain and receive the message. When dealing with other professionals in business relationships, this is not always as simple as it seems. There are plenty of opportunities for misunderstanding and miscommunication.

> **Networking hint**
> What is communication? In short, it's signalling. It's the transmission, by speaking, writing or gestures, of information that evokes understanding.

6.1 What happens when you open your mouth?

If you manage to insert both feet with speed and agility, you're probably just a bit nervous. Don't be surprised if words come out that you seemingly have no ability to control. A conversation under these circumstances can go seriously wrong before you've had time to do much more than sit down.

There are some points to remember when considering the various methods of communication and some hazards to be aware of when dealing with business relationships.

Only 7 per cent of the impact you make comes from the words you speak and that 7 per cent comprises:

- the type of words you use;
- the style of sentences you use;
- how you phrase them.

> **Networking hint**
> If you want to make a favourable impression with your team, consider the words, the ideas and structure of the message you wish to convey. Keep it as simple as you possibly can. Aim for clarity over ambiguity.

Let's look in a little more detail.

- Commonly used words, in short direct sentences, have the greatest impact and allow the least margin for error or misinterpretation.
- Long words wrapped in complex sentences are confusing and best avoided. Don't use jargon, either, unless you are sure it will be understood by all those present.
- Positive statements are far more acceptable and will gain you greater advantage than negatively expressed remarks.

6.2 Using your voice effectively

Pay attention to your voice, too. Tone, inflection, volume and pitch are all areas to consider. Relatively few people actually need to develop their speaking voices, but many do not understand how to use their voices effectively. The simplest way is to compare the voice to a piece of music, because it is the voice that interprets the spoken word.

Those who have had some training in public speaking sometimes use mnemonics as memory joggers for optimum vocal effect. One simple example is RSVPPP:

Rhythm
Speed
Voice
Pitch
Pause
Projection

6.2.1 Rhythm

Speaking without tonal variety can anaesthetize your listener. Try raising and lowering the voice to bring vocal sound to life (and keep your audience awake). Rhythm is directly linked with speed.

6.2.2 Speed

Speed variation is connected to the vocal rhythm. Varying speed makes for interested listeners and helps them maintain concentration. If you're recounting a story, speed helps to add excitement to the tale. But the speed of delivery should be matched with the volume you're speaking at.

6.2.3 Volume

Level of volume obviously depends on where the conversation is taking place. It would be inappropriate to use loud volume when speaking in a one-to-one situation. However, you'd probably need to increase it if you were talking in a crowded venue, such as a business reception or work area. Volume is used mainly for emphasis and to command attention; lowering your voice can add authority when telling an interesting story or giving advice.

6.2.4 Pitch

Pitching your voice is something public speakers do. They are trained to 'throw' their voices so they can deliver their speech clearly to their audience in whatever size or shape of room they're speaking in. In general, it's irritating to any listener if they have to strain to hear what the speaker is saying.

In normal conversations where you need to be heard clearly (for example, in restaurants where there is continuous background noise as well as the hubbub of other voices), it's impossible to pitch your voice if you hardly open your mouth to let the words out. Correct use of mouth, jaw and lip muscles will produce properly accentuated words and assist with clear enunciation. Pay attention to these facial muscles otherwise your voice will be just a dull monotone.

6.2.5 Pause

Practise the pause. It can be the most effective use of your voice though it is often ignored. A pause should last about four seconds. It sounds like an eternity, perhaps, but anything shorter will go unnoticed by your listener. You can use the time to maintain good eye contact. The effect can be dynamite. Remember the 'er' count. Filling spaces in conversation with props such as 'ers', 'ums' or 'you knows' where there should be pauses is a clear sign of nervousness.

6.2.6 Projection

This encompasses everything about the way you come across: power, personality, weight, authority, and expertise – what some people call 'clout'. You want to build some powerful professional connections. It pays to have gravitas in your dealings with people. Projection is an art that can be practised. But you can learn so much from listening to experienced communicators who are experts in these skills.

6.3 Reviewing your vocal skills

If possible, get a colleague or a friend to give you feedback on your voice and mannerisms. Unless you get an accurate appraisal, you could be spoiling your chances of successful exchanges. With practice you'll be surprised how quickly some of these traits can be eradicated. Once you've eliminated them and developed some of the skills suggested here, the improvement in your style of conversation and self-confidence when meeting people will be remarkable.

- Be clear – use simple, easily understood words and phrases.
- Be loud (enough) so that your listeners can hear you.
- Be assertive – a bright and confident tone will inject interest into anything you're saying.
- Do stop for breath. It's essential to let your listener digest what you've said – and to have the opportunity to respond.

> **Networking hint**
> Remember, your voice is an instrument, just like your body. It is also, like your body, very flexible. You know the expression, 'It's not what you say, it's the way that you say it.' That couldn't be more true.

6.4 Face-to-face encounters

The key to success is to get onto the other parties' wavelengths as soon as possible. By putting yourself into their shoes you'll demonstrate your ability to *empathize* with them. They'll find communicating with you easy and respond positively.

One of the most important aspects of communicating is to develop good *listening skills*. Lots of people are not good listeners. You are not alone if you are far more interested in what you have to say than what the other people are saying to you. Poor listening damages exchanges and that is what you are at pains to avoid. Good listening avoids misunderstandings and the errors that result from them. We'll now look in more detail at what the behaviour of a good listener should be.

6.4.1 Listen carefully

A person who is listening attentively keeps a comfortable level of eye contact and has an open and relaxed but alert pose. You should turn towards the speaker and respond to what she is saying with appropriate facial expressions, offering encouragement with a nod or a smile.

6.4.2 Be disciplined

Adopting the behaviour of a good listener will help you establish good rapport with the others. It requires a degree of self-discipline and a genuine desire to take on board the message the speaker is trying to convey. You need to be able to suspend judgement and avoid contradicting or interrupting them. Postpone saying your bit until you are sure she has finished and you have understood their point.

6.4.3 Be reflective

Reflecting and summarising – repeating back a key word or phrase the speaker has used – shows you have listened and understood. Summarising gives the speaker a chance to add to or amend your understanding. Your colleagues are far more likely to listen to you if you let them know that you have heard what they've said by using the tactics of reflecting and summarising.

6.4.4 Pitfalls to avoid

Pitfalls to avoid include thinking up clever counter-arguments before the speaker has finished making their point. Don't interrupt unnecessarily or react emotionally to anything that is said. If the subject becomes dull or complex, don't register your uninterest by succumbing to distractions or fidgeting.

6.5 The five levels of listening skills

6.5.1 Level 1

The first and worst level is **ignoring the speaker**.

You look away, avoid eye contact and do something else altogether. This is dreadful in a business context. Your colleagues will never give you the time of day again if you commit this cardinal sin.

6.5.2 Level 2

The second level, which is almost as bad, is **pretending to listen**.

In some ways this can be quite dangerous. If you're nodding your head, and saying 'mm, yes, aha' when you actually have no idea what's being said, you could be in for a nasty shock. Don't be surprised if you hear your colleague saying, 'So you'll run in the London Marathon next year on behalf of my favourite charity – how wonderful.'

6.5.3 Level 3

The third level listening skill is **being selective**.

You may well find yourself listening for key words that are of importance, such as 'successful tendering' 'Phase One completion' 'on schedule'. The result is that you could miss the context of the exchange. Your colleague could have been telling you that the project is not going ahead as planned or is delayed for some reason.

6.5.4 Level 4

If you can develop the fourth level skill, you're doing well. This is called **attentive listening**.

You are focused, with positive body language, leaning forward, nodding your head appropriately and maintaining eye contact. The other members of the group know you're paying attention and this creates an atmosphere where they'll want to share valuable information and engage in serious dialogue.

6.5.5 Level 5

The final level is **empathy**.

Empathy is the ability to put yourself in someone else's place and see things from their perspective. This takes time to achieve but it will impress anyone once you have reached it. It is the art of being able to identify mentally and emotionally with your communicator, fully comprehending the tones, pitch, body language and other subtle messages your fellow interlocutor is conveying.

It is totally exhausting to do this for any length of time but it will take your professional relationship to a much higher level rapidly. You will have included each other in the closest of possible personal networks (sometimes called the *virtual team*). He will consider you one of his first ports of call when information gathering or project awarding is required, and you'll willingly reciprocate.

6.6 Directing the communication cycle

Can you recall a time when you've been chatting to a colleague and you've looked at your watch and said, 'Wow, is that the time? I must have been talking to you for ages.' This usually happens when the two people concerned are giving each other space in their conversation. There is a feeling of ease, ideas are being passed to and fro, and a natural exchange develops. It's a bit like having a conversational game of table tennis. This is called **rapport**.

> **Networking hint**
> If you can begin the rapport-building process with your professional colleagues, you will begin to cultivate the relationship you want. Your exchanges will become frequent and more valuable. You're attempting to establish the balance of listening and talking.

There are times when you'll want to find out more information. It's easy to ask too many questions and fall into a sort of Spanish Inquisition situation. Conversely, when responding to a question, you can give away too much information. If you're on the receiving end of this from your colleagues, the relationship may not make much progress. No one likes to feel they are being 'pumped' for information. It's infuriating, rather insulting and you'll want to distance yourself as quickly as possible.

Only one person at a time can truly direct a conversation. One leads and the other tends to follow. This doesn't mean there is no give and take. Neither does it mean that the other party is subservient. But one of the parties should lead and there is merit in your being the one who does so, if your objective is to build a proactive professional relationship with rewards for both sides.

6.7 Opening rituals

At the start of a meeting, there are usually some general opening remarks. This sort of ritual is customary and should take no longer than a few minutes at the outset of proceedings. Watch for the moment when the chattiness should cease, because, if you have no real plan, the other members of the team may lead you off into uncharted waters. Then you may find yourself heading in the wrong direction.

Someone usually starts off by saying, 'Right, shall we move on? Can you tell me ...' That person could be you. If no one else seizes the opportunity to take control at this point, you should, or else you may have lost the initiative for the rest of the meeting.

You might consider going into the meeting with a short agenda. If this isn't written down, at least it should be in your head. It could be little more than a few helpful suggestions. Perhaps you've already aired the topics for discussion in a telephone call beforehand. There is no rule here, but whatever has been agreed it does at least mean that the exchange proceeds along some agreed lines.

It also provides an element of control during the dialogue, if the conversation meanders into other areas. You could refer back to your brief by saying something like, 'We were going to discuss X next.' And then move on smoothly to the next stage. The early part of any meeting is a key stage for your confidence. You'll feel and operate better if you get off to a planned start and you'll be able to maintain better control and direct the rest of the exchange.

6.8 Good conversational techniques

To develop a balanced style of communication, try to begin the conversation by introducing yourself and giving some personal information, your position and something about the brief.

This is called the **inform** stage. Once you've given this, ask a direct question of the other members of your project team. This is called the **invite** stage. Then **wait** for the response. On receiving this, listen to every word. Then **acknowledge** and, if necessary, repeat the essence of the response.

If you achieve this cycle of communication you can repeat it many times over during the encounter to establish a good rapport between you and the other parties. It should make the time pass effortlessly and harmoniously and make your exchanges a pleasant experience.

Building good relationships with other members of the project team means you are attempting to get closer to your team members by developing the art of good conversation, so pay attention to the importance of **eye contact**. Appropriate eye contact at all times in the exchange is essential. If you are talking, check that the rest of the group are paying attention. They should be looking at you with an interested expression, nodding occasionally and smiling at the right times with an alert and open posture.

6.8.1 Things to look out for

However, should one of your colleagues appear to be **falling asleep** during one of your conversational gambits, it could mean that: he's had a late night; he's had an early start; he's suffering from jetlag; the atmosphere in the room is too stuffy; or your dialogue is rather boring. Don't wait until his head falls forward and hits the desk. If you fail to notice until you hear the crash, you're definitely talking too much.

Keep an eye out, too, for **fidgeting**, since this could indicate that: you've lost his attention; he wants a break; he's irritated by something you've said or he finds the conversation irrelevant.

Whatever the reason, it's time to shut up. Close mouth without delay and smile. Hopefully, with a bit of silence you can retrieve a conversation that may have got off to a rather inauspicious start.

Should one of your colleagues start **shaking her head**, this could mean: she wants to say something; she doesn't agree with you; or she simply hasn't a clue what you're waffling on about. Again, as above, time to bring your remarks to a swift close. If you think you've **lost her attention** completely, and she has turned off, try to regain it by asking her a pertinent question. Re-establish eye contact and vary the volume or expression in your voice.

6.9 Other forms of communication

6.9.1 Telephone calls

Telephone calls can be difficult to deal with and can often cause trouble between parties who do not know each other all that well.

First, because you can't see each other face to face, you have to rely on tone of voice. This can be deceptive. He may sound uninterested because he's talking in a low voice. It may be something as simple as the fact that he's got a sore throat, or he's trying to avoid the rest of the office hearing his conversation.

It's essential to pay attention when a member of your new project team calls. If he's on a mobile, you may well get a distortion, due to background noise, traffic, airport announcements or similar. If possible take the phone call in a private place so as to avoid even more noise coming from your end of the phone.

6.9.2 Voicemail messages

There's an art to leaving successful voicemail messages. It's simply this: be clear and be concise. Don't speak too fast. If you are leaving your telephone number, slow down. Speak slowly while recording the information.

If the message you leave is either gabbled or garbled, it will be impossible for anyone to return your call. It helps to leave a date and time when you record your message, so that the other party can respond quickly if time is critical.

6.9.3 Text messages

This is the perfect form of communication for quick exchanges of information. One word of warning: don't use confusing abbreviations. If you received the following message – 'CU 7.30' – does that mean 'See you at 7.30 p.m.' or 'Curtain up at 7.30 p.m.'? Check if you are in doubt.

6.9.4 Written communication

The main point about written communication is that, whatever form it takes, the recipient cannot see you or hear you. Your colleague has no option but to accept what she reads. You should pay particular attention to wording and expressions, because, if it is at all ambiguous, it is liable to be misinterpreted.

6.9.5 Letters

When handwriting letters put yourself in the position of your recipient. Write neatly and clearly and make sure your spelling is correct. It helps to use a decent pen and good-quality paper. Impressions count, remember.

With a personal thank-you note, use the recipient's business address because it is after all a business relationship, even though you are thanking the person for inviting you to a social occasion. Keep your message simple and make it easy to read.

6.9.6 Email

Much has been written about email etiquette, because this is such a popular and efficient form of communication.

If you wish to email to your colleagues, check that you have their correct email addresses. There may be confidentiality issues – particularly if your exchanges include something other than the job in hand. If you are using a business email address, do be circumspect. Emails may not always reach the recipient directly. Some people have staff who scan emails before forwarding them to the main addressee. Consider the likelihood that your email is going to be read by someone else, and be careful.

On a more practical level, email is not the medium for rambling on and on about the project, or any other subject that is dear to your heart. Keep email communication clear and short. It's no substitute for face-to-face contact, but it does allow for a fast exchange of information, particularly when confirming meetings or referring to matters just discussed.

6.10 Communication-skills awareness checklist

Presence: Pay attention to the way your voice and body language are used in conjunction with the words you speak. You can convey the right impression if they are used correctly.

Relating: Don't underestimate the importance of developing your rapport-building skills to get on the same wavelength of your professional colleagues.

Questioning: When engaged in conversation with the other parties, make sure you match your question to the situation or subject. Beware of asking irrelevant questions: this will show that you've not paid attention to what was said.

Listening: Listen to everything the other person says attentively. Try to reach at least 'Level 4: attentive listening (see Subsection 6.5.4 above). If he's likely to become a significant influence in your business-development strategy, you should aim for Level 5 eventually.

Checking: This is the art of glancing at your colleagues to see that they are still on your wavelength while you're engaged in dialogue. Watch for gestures and see whether they do the same when they are talking.

6.11 Manners and mannerisms

Politeness among business colleagues is always desirable but sometimes lacking. Being courteous to colleagues, staff and clients will significantly influence relationships in the workplace and foster a harmonious culture. Hopefully, you are not a troublesome character, but go out of your way to deal pleasantly with others. It is important to preserve other people's dignity and respect, particularly if you are keen to develop good rapport-building skills. The colleague who rides roughshod over other people's feelings in a team (however well she is regarded professionally) does often end up with a load of headaches – mainly caused by her own actions.

6.12 Non-verbal communication

If you are familiar with the practice of non-verbal communication it can be used effectively to SOFTEN the hardline position of others. Consider this acrostic:

Smile
Open posture
Forward-looking
Touch
Eye contact
Nod

If you approach someone with a smiling face, it encourages a similar response from them. Just as important is your presentation and body language. The posture should be open with head upright, and you should stand straight but with hands relaxed by your sides. Make appropriate gestures to show that you are welcoming the exchange

Eye contact should at all times be honest and open. Avoid staring but maintain a steady gaze when speaking to colleagues. At the same time, encouragement by nodding your head shows consensus and indicates that you are taking note of the points being made.

Polite behaviour should ensure smooth interaction between colleagues working on a team or project. If you practise giving the right signals, with luck, others will mirror your actions. After all, imitation is the sincerest form of flattery. If for some reason the project team is full of people who are backstabbing and politicking, this could spell difficulty ahead. Sometimes a working atmosphere worsens because of a new arrival on the team. Some people do seem to act as virus spreaders – those whose presence brings a chill factor that spreads like an epidemic.

Maybe one of your colleagues' approaches needs a makeover? It is easier to adapt an individual's manner when dealing with others than it is to alter an endemic organizational culture. If the problem is not too great to be rectified, a little praise goes a long way towards helping with these issues.

6.13 Maintaining good morale

Good traditional praise does not go amiss in the workplace. One of the keys to retaining goodwill among professional colleagues is to foster a sense of camaraderie among them. The more control team colleagues have over how, when and where their work is done, the happier they will be. Their performance improves, along with their morale. There is less confrontational behaviour – in short, everyone gains.

Regardless of how advanced technology has become and what the latest equipment and gadgets enable us to do, some things are in danger of becoming too impersonal and remote. There is nothing more encouraging than good manners and personal attention. Where a plague of bad behaviour or ill manners pervades the workplace, there is bound to be an increase in problems among colleagues.

Kindness should not be underestimated. If someone is showing signs of anxiety, stress or depression, they are probably feeling inadequate and undervalued. Left unchecked, this situation could spiral towards absenteeism. Work-related anxiety has knock-on effects because it doesn't go away if you ignore it. There is a marked difference between stress and anxiety: the former is caused by overstimulation and overload; the latter is usually because the person feels a failure, a poor performer or inadequate. Should there be any signs of bullying, harassment or sexual or racial prejudice towards someone you work with, these could result in a number of physical symptoms: shortness of breath, nervous behaviour, backache or headache, eating disorders and insomnia.

Although dealing with anxiety is best done by seeking professional help, starting with a little kindness can go a long way to combat the effects. Helping colleagues cope with challenging situations by praising them can alleviate the anxiety symptoms. However severe the anxiety displayed is, there is no doubt that being kind to others will help to redress the balance. One important tip is to note how people breathe. Anyone showing nervousness, anxiety or stress, will hold their shoulders in a rigid way and have shallow breathing. Think about any relaxation techniques you may have been taught. They all encourage deep breathing, holding your shoulders down and holding your arms loosely at your sides. If you are able to observe your colleagues' breathing and posture, it should be quite evident who is not relaxed or at ease. Start your kindness campaign with them.

If you are expected to deliver bad news – some criticism to someone who is underperforming – it is still possible to do so without hurting someone's feelings. Sweeten your message by making it clear that you are trying to help them, and work out what it is you want them to do differently. Have the conversation with them at an appropriate time and place, so as to avoid any embarrassment to them in public, particularly if it is a sensitive issue.

> **Networking hint**
> In sensitive issues, follow the maxim of Catherine the Great of Russia who once, said, 'I praise loudly. I blame softly.'

You need to be sure that the person knows what you are talking about – whether it is their performance over the past few months or a piece of work they have delivered to you that morning. Be specific and keep it concise: 'Your poor timekeeping's causing a problem for the department' is precise. Reminding them that they have been a good role model in the past will reinforce their identity and encourage them to think positively about themselves.

Professional colleagues whose motivation levels are high are not likely to be suffering stress or a lack of self-esteem. Should grumpiness pervade the workplace, pay careful attention. If you are able to dispense honey rather than vinegar, you might influence your colleagues towards a change of mood. Smiling is the first step, and being happy around them should help to encourage a warmer atmosphere. Beware the chill factor: should moodiness be allowed to prevail, you will find this contaminates the team fairly swiftly. The benefits of working with good-tempered colleagues are high. Befriend the difficult if you can. It makes it much harder for them to be unpleasant if you're nice to them.

It is curious to note that, despite the many forms of communication available today, if you receive a personal note from someone, with a hand-addressed envelope, you are almost certain to feel good about it. It is rare to get handwritten messages and what it conveys is that you – the recipient – is worthy of respect. Should the message contain an unsolicited thank-you or testimonial for something you've done, you'll probably be walking on cloud nine for the rest of the day.

One of the easiest ways to disarm the opposition is to be charming and to smile. Many people have wonderful natural smiles, but, due to nerves or apprehension, their faces frequently set in serious expressions. It takes far fewer face muscles to smile than it does to frown. If you are going to praise someone for something they have done, you really should be smiling when you speak to them.

A sign of a confident person is someone who, while speaking in public, or giving a performance, allows themselves to project a warm encouraging expression. When a smile lights up your face, people will notice you. It is quite likely that their natural reaction will be to smile back. Imagine the powerful advantage this gives you in your potentially awkward situations. People who smile give the impression of being pleasant, attractive, sincere and confident. It relaxes those around you, with whom you are about to communicate.

Charm, good manners, politeness – these are all somewhat traditional standards. Do you notice a person's eyes when you are with them? Do you find that their attention remains focused on you throughout the exchange? Or do their eyes stray when someone walks past the door, or if a commotion takes place outside? Keeping your eyes and ears directed towards the person you're with is vital if you are trying to placate them or restore calm to a difficult situation. By creating the impression that your attention is all theirs, you will have a strong effect on the bridge-building process in your area of concern or conflict. All these things build confidence. A conflict can be resolved or a relationship enhanced purely through a display of confidence. Self-belief and self-assurance are vital if you are to realize your potential and maximize your success at dealing with challenging individuals.

One of the nicest things about praise is that it stops the majority of people complaining. Whingeing is becoming increasingly common at work. If there are one or two Olympic-standard moaners among your professional colleagues they need to be managed properly to move them away from their aggrieved attitude. Simply admonishing them and suggesting they 'keep a stiff upper lip and get on with it' won't work. Frustrated colleagues are not always wrong or being difficult. Sometimes there are fundamental problems and these should be investigated swiftly by whoever is in charge of the team.

Negative perspectives can be dangerous and, if there are a number of people making waves within a team, this is where the rot can set in. You cannot afford to have a vicious circle of doom-laden prophets fuelling their own gloomy forebodings. This sort of enemy within is insidious and can, like a virus, lie in wait until another unsuspecting prey comes along ready to be infected. One of the most effective antidotes is praise and reward of even the smallest amount of success or progress. This will inevitably lead to bigger things. Why not, for example, keep a 'success diary', or chart, so that team members know when they have contributed to a mention in the book or on the wall chart.

6.14 Dealing with professional envy

There is a huge amount of professional envy reported in the workplace. It is an issue that can badly destabilise a project team and requires careful handling. You can't simply charm someone while they are suffering a particularly violent attack of the green-eyed monster. If you're doing really well someone could be experiencing envy and resentment about you. Success sometimes spawns this type of jealousy among professional colleagues. One reason for professional jealousy is that of feeling threatened. Is someone feeling resentful towards you because of your recent promotion? Or are you popular and polished and they not?

Some people regard work as a competition with only one winner. Should you be seen to be edging closer to the chequered flag, they may see you as taking the prize and themselves losing. The only way they can deal with this is to push themselves forward and hold you back at the same time. Job jealousy can take many forms: from the odd 'forgotten' message or 'mislaid' written instruction to a formal request for information that fails to get delivered. Should you feel that your work is being undermined or criticised unfairly you need to make a note of such things. If the behaviour becomes personal – remarks about your appearance, for instance, or an aspect of your private life – you should make a move to counter it.

Developing a thick skin helps, as this will afford some protection from your workplace monster. Focus on what is important and ignore what is trivial. Don't allow yourself to show pride openly in your promotion at work. If you want to bask in self-congratulation, make sure you do it out of the workplace among friends. You need to maintain your own standards, rather than dropping them to the envious colleague's level. This will do you a lot of good and you will be confident you are not contributing to making the situation worse.

6.14.1 Tackling the situation

You may have to be brave and tackle the person out in the open. Confront them and ask if they have a moment, because you have something on your mind that needs to be discussed. You could say it won't take long. Your request for a chat will probably be refused, in which case ask whether this is just a bad time, and would they suggest a better time? Don't be put off by them. Persist until they agree to talk to you.

You could open the conversation by saying that you feel there is a bit of tension between you and you wonder if everything is all right. If they reply along the lines of, 'I've no idea what you're talking about', then you can say, 'Oh, good, I must have been mistaken. I'm glad, because I want us to be able to work together.' Make it clear that, should they have something they wish to discuss in future, they must say so. It is important that the exchange conclude on a positive note, where both parties can exit with some dignity.

Should matters not improve, you will have to face another one-to-one, at which you become more assertive. You will need to spell out the point of grievance: 'It is not acceptable when you do . . . I find it difficult working with you. Please make sure that in future you do . . . Are you OK with this?' Unfortunately, if this doesn't work, you have no choice but to make a complaint along formal lines. Disciplinary action may follow, but, since there is little chance of your establishing trust and rapport with this colleague, there is nothing lost.

Whatever you do, keep every exchange as professional as possible. The most important thing is to try to work through this difficulty without raising problems with others in the department.

6.15 Communicating with confidence

Two of the most important skills to acquire for dealing with other professional colleagues are confidence and assertiveness. Whoever you are working with, the ability to conduct yourself appropriately with other professionals is highly desirable. This includes creating the right impression, building up your confidence and self-esteem so that you can manage relationships with high-maintenance and potentially difficult people more easily and successfully.

Networking hint
Building up your self-assurance is very important. Confidence – like a muscle – needs to be exercised if it is to develop.

6.16 Visual impressions

Visual impressions are as important as oral messages. You are working in a professional environment, so, if you sense there are likely to be personality issues, it is helpful to contain any potentially difficult situations before they get out of hand. It is an enormous advantage if you can remain calm and in control, so that, even if others are being provoking, you can take steps to limit the trouble they may cause.

Getting off to a positive start makes things easier. A good beginning not only affects the quality of the encounter, it affects your confidence too. Confidence requires preparation and needs to be actively worked at to ensure you achieve the right impact. If you are well prepared, mentally and physically, for an important team meeting, you will appear more confident, calmer and better able to handle any difficult situations that may follow.

This is not a question of tricks or gimmicks. It's about being businesslike and professional and aware of the importance of having everything go well in the early stages. Having the intention is the first step towards achieving it.

6.16.1 Looking the part

If you want to be seen as confident and self-assured among other colleagues, capable of diffusing difficult business situations in a cool professional manner, using open body language will make you more persuasive. Remember, your body is an instrument – it conveys every emotion. A good tip is mirroring gestures. These are great for creating a good first impression with a challenging person. Copying what the other person does sends a positive message, endorsing what information they are conveying.

If you want to create a favourable impression with someone, your body will quite naturally point towards them – your face, hands, arms, feet and legs. These gestures can be subconscious, but the other person picks them up quite easily. Have a good look next time you're in a group of people – whether a social or work situation. Observe how individuals position themselves when communicating with each other. They naturally angle themselves towards the person with whom they are trying to create a positive impression, and turn away from those they are seeking to avoid.

6.16.2 Eye contact again

Making the correct sort of eye contact in work situations is important. You are probably dealing with someone you don't know very well, so there are a number of things to remember. It is natural to look at people from eye to eye and across the top of the nose. This is the safe area to which eye contact should be confined. With friends, away from work, this area of vision increases to include both eyes but also downwards to the nose and mouth. If you're nervous, you should avoid staring at someone when they're speaking. On the other hand, looking away completely, blinking or closing the eyes for longer periods than normal can indicate shyness – an attempt to block the situation.

In a meeting when conveying information, pointing to something you are discussing (such as a model or drawing) directs attention towards it, and away from you, if you need a break from all eyes being focused on you. You can then bring the focus back again by lifting your head and engaging eye contact again. This is helpful if you need to change the emphasis of your meeting.

> **Networking hint**
> One important point from Chapter 5 is worth repeating here: you've been given two ears and only one mouth, so use them in that proportion.

Bear in mind that, if you spend twice as much time listening as talking, you will create a positive impression. Other people will regard you as a skilled communicator who can operate effectively in potentially awkward situations.

6.17 Good manners

How do you feel if someone bothers to say thank you if you've travelled some distance to see them, or made an effort to attend a meeting when you are very busy? When dealing with other professionals who also have hectic work schedules, a businesslike approach will stand you in good stead.

Turn up on time when you have an appointment. The ability to be punctual at a first meeting gives the overriding impression that you are well organized and capable of delivering (even if it is just yourself). Turn up late, however, and all that will be remembered is that you missed the appointment. It may sound harsh, but this could jeopardise a future working relationship. A lot of hard work will have to be done to help redress this, so don't make things difficult by an overly casual approach. However organized you are, you should always allow extra time when travelling, to avoid stress. If you arrive for a meeting in a fluster and out of breath, you'll be in the wrong frame of mind and won't be in control. Appearing cool, calm and collected is well worth the extra investment of getting out of bed an hour earlier.

6.18 Pay attention

This may sound like unnecessary advice, but it is surprising how many people can't stop their eyes straying when someone walks past an office or a commotion takes place outside. Keeping your eyes and ears directed towards your colleague, or whoever is speaking, is vital if you are in an important meeting where temperaments are clashing. Don't relax – show that you are giving the situation the attention it deserves. Showing that you are concentrating will create confidence in your ability to cope. This is particularly important in meetings where a number of characters are competing for the upper hand.

6.19 Switching off the mobile

There's no better way to irritate people in a business meeting than being interrupted by an unwanted bleeping coming from your pocket or bag. Despite constant reminders, people still forget to switch off their mobiles and it never fails to annoy others. Don't compound the sin by answering your phone. This rule applies the other way round too. If one of your colleagues or contacts has the insensitivity to receive calls and messages throughout a meeting, it's an insult. It shows a lack of respect not only for the occasion but also for the others present, and creates completely the wrong impression.

There are occasions when such interruptions are unavoidable – for instance, if you are awaiting the final details of a report, or the outcome of an enquiry. If you have to conduct a meeting while expecting an important call, have the courtesy to say so and announce the reason why you are awaiting some information. Then, when the phone call does interrupt the dialogue, there is no need for embarrassment. Just excuse yourself for a moment while you take the call, make a discreet exit and be brief.

6.20 Confidence boosting

If you are keen to establish good relationships with colleagues, some of whom can be highly competitive, you should have a plan for boosting your morale. With healthy self-esteem you will have the confidence required to work closely with challenging individuals in the most favourable and positive way. Building yourself up so that you believe in your ability to succeed is very important. Behave and look as if you had already achieved your goals, and you are halfway there. Confidence breeds confidence, and, as it develops, it will become natural to you and have a positive impact on others.

It doesn't matter who you are, people make judgements based on their first impressions. One of the key reasons why you should spend time and effort in preparation, both mental and physical, for challenging meetings is to give yourself an advantage in difficult situations. If you can practise this, you will find many situations far less threatening. The outcome of conflict situations is often determined by the confidence shown by the parties involved. A lack of skill or knowledge can go unnoticed if you have self-assurance. A conflict can be resolved and the respect of fellow professionals earned through a display of confidence.

Self-belief and self-assurance are vital if you are to realize your potential and maximize your success at dealing with other professional colleagues.

Chapter 7
Managing other people

Managing people is something that you may have to do, if you're not doing it already. If you are working on improving your interpersonal skills, you will become a better manager as a result.

You may have originally been appointed because of a different set of skills – scientific, technical or other. Should you be technically excellent at your job and have management capabilities, you will almost certainly be promoted. At some stage your original skills become to a large extent redundant, because what you are using most of the time now is management skills.

But what if you haven't got those skills? This could have dramatic consequences for your career plans. Some people in this situation are fortunate enough to be offered management training, but the vast majority are not. You have no choice but to muddle through, or you could copy other people or emulate a role model, all of whom may have had to do precisely that in their own past.

This chapter tackles two of the foundation stones of effective management practice – delegation and appraisals – both of which are critical to developing and improving performance in others. They require advanced relationship-building skills within the organization.

7.1 Delegation

'Delegation' is a word that is bandied about liberally and is something to which people pay lip service. People rarely delegate effectively. Mostly what happens is that you get told to do a job or take something on – that is not delegation. Very often it is an either/or situation. Either you get dumped with something you can't cope with or you don't get a chance to prove your worth because delegation is not implemented effectively. More often than not it hasn't been thought through, which results in things going wrong, breakdowns, upsets and so on. All very demotivating for staff and project managers alike.

As a manager you will have to assign or allocate work to others in your team(s). This will be done by balancing the work that has to be done against the availability of the other people and their abilities. Some of the work may be routine and repetitive; some may not. When you assign work to a team member, you may retain the decision-making responsibility if it becomes necessary to decide upon an alternative course of action. Delegation goes one step further and implies that the authority to make decisions is given to the team member.

> **Networking hint**
> If you cannot delegate effectively you will find your own development will suffer and you will become snowed under with work. You need to recognize the importance of delegating work to others in your team so that you too can develop and grow.

7.1.1 Some reasons not to delegate

There are several reasons why some managers feel reluctant to delegate. One of the most frequent excuses is, 'It's easier to do it myself.' That may be true to start with but it soon becomes a vicious circle: the more you have to do, the easier it is to do it all yourself because it is quicker than taking the time to delegate. But that road leads to overload for you and loss of morale for others. Ask yourself what might be preventing you from delegating; perhaps it is because you:

- do not understand the need to delegate;
- lack the confidence with team members, and therefore will not give them the authority for decision making;
- do not know how to delegate effectively;
- have tried to delegate in the past, but failed and so will not try again;
- like doing a particular job that should be delegated, but will not delegate it even though you know the team member would enjoy the job;
- do not understand the management role or how to go about it;
- are frightened of making yourself dispensable, so keep hold of every job;
- have no time to delegate;
- have nobody to delegate to.

All of these barriers need to be overcome if you are to delegate effectively.

7.1.2 The skill of delegating

Delegation is a skill that, like any other, can quickly be learned. Most of it is common sense, but here are some tips for effective delegation.

- Plan delegation well in advance.
- Think through exactly what you want done. Define a precise aim.
- Consider the degree of guidance and support needed by delegate.
- Pitch the briefing appropriately. Check understanding.
- Establish review dates. Check understanding.
- Establish a 'buffer' period at the end, in which failings can be put right.
- Delegate 'whole jobs' wherever possible, rather than bits and pieces.
- Inform others involved.
- Having delegated, stand back. Do not 'hover'.
- Recognize that work may not be done exactly as you would have done it.

- Do not 'nit-pick'.
- Delegate, not abdicate, responsibility.

7.1.3 What should be delegated

A manager must analyse the job they are actually doing in order to establish what can and cannot be delegated. You need to identify:

- totally unnecessary tasks that need not be done at all;
- work that should be done by another person or in another department;
- time-consuming tasks not entailing much decision making, which, provided training is given, could be done as well by the team member as by the manager;
- repetitive tasks that over a period take up a considerable amount of time, but require more decision making and would serve to help develop a team member.

A delegation plan and timetable must then be proposed to enable time to be found to delegate. Except for the simplest of jobs, you will find that something like eight to twelve times longer will be needed to delegate a job effectively than actually to do it. However, by taking the time to delegate properly in the first place you will save yourself far more time in the future. See it as an investment in your own future as well as in the future development of the delegate.

7.1.4 What should not be delegated

There are always certain tasks and authority that a manager should not delegate. This does not mean that you cannot employ staff to assist with these areas of work, but you must remain the final decision maker. These areas of work are:

- being forward-looking and constantly seeking opportunities for the enterprise;
- setting aims and objectives;
- creating high-achievement plans for your department or the part of it for which you are responsible, and ensuring that quality standards are developed and maintained;
- coordinating activity – that is knowing the task that has to be done, the abilities and needs of your own people, the resources available and then blending them to achieve optimum results;
- communicating with your people and with senior managers and other professionals and colleagues;
- providing leadership and positive motivation;
- the training and development of your project team;
- monitoring and surveying everything that is going on and taking the necessary action to maintain the planned level of achievement and quality performance.

From the above, it is clear that you will not weaken your position by delegating work that does not fall into any of these areas. In fact, the contrary will be true. You will free yourself to do the jobs that you alone can do, and should do, to be effective.

7.1.5 How to delegate

Once a delegation plan has been prepared, each job must be taken separately. You must then prepare a specification that will state:

- the objective or intended goal of the job;
- the method you have developed to do it;
- data requirements and where the information comes from;
- any aids or equipment needed to do the work;
- the principal categories of decisions that have to be made;
- any limitations on authority given to make these decisions, such as when you should be consulted.

When this preliminary specification has been prepared, you must start training the delegate to do the job. Initially, close control should be maintained, but this should be loosened as soon as possible. Some form of control must be maintained, but this should not be more than is necessary to ensure that the job continues to be done properly.

> **Networking hint**
> Keep track of which jobs you have delegated and to whom. Monitor the process with each delegate from a tactful distance.

7.1.6 Advantages to project team members

Delegation is often seen as being of advantage to you, the manager, but it is also of considerable benefit to the team members concerned. The fact that jobs you have developed are passed to others to do, along with the requisite authority to act, is an aid to the development of individuals both practically and psychologically.

7.1.7 Delegation exercise

As you read through the list below, tick any items that you feel particularly apply to you. Then consider whether the suggested changes in what you do might be helpful.

Barriers to delegating	How you might tackle the barrier
I find it difficult to ask people to do things.	Try explaining to them what you will be freed to do if they take on the task
I do not have time to delegate.	Decide to break out of the vicious circle and make time. By investing half an hour explaining the task, you save the three hours it will take to do the task.

It is quicker to do the job myself; explaining it to someone else takes too much time.	It may be quicker to do the job yourself, but you have a responsibility to develop your staff members' skills. You will get quicker with practice.
I could do the job better.	Being responsible for developing the skills of your staff means investing time in development and training. In the long run it will save time. Set up a development programme.
I need to know exactly what is happening.	As a manager, you must get results through other people, or you will become overloaded – so you need to trust your staff. Build in regular feedback.
I enjoy this job. I've always done it.	As a manager, you have to let go of tasks that other staff can do. Do only what you can and should do.
I am afraid it won't get done properly, and I will get the blame.	Prepare for delegation, and build in controls as the job is done. You have the right to make mistakes.
I am afraid someone else will do it better than I can.	Set targets for your team members to do better than you at specific tasks.
Someone else will not do it my way.	Agree to goals and targets and give freedom. There are often many ways of doing a job. A good team benefits from a variety of approaches.
I am not sure how to do this task so feel I had better do it myself.	You need to decide how to tackle the task before deciding whether it's suitable to delegate.
The job is too big/important.	Break the job down. Most jobs contain some routine elements which can be delegated.

7.1.8 A good delegator or a willing martyr?

Delegation is one of the most difficult things that busy professionals need to learn. Take the test below to see what kind of delegator you are.

1. What does delegation mean to you?
    a) Passing the buck to juniors?
    b) Dumping responsibilities?
    c) Tricking others into doing work that is rightfully yours?
    d) None of the above?

2. Are you nervous about delegating because:
    a) You do not trust anyone else to do the work?
    b) You do not want to overburden someone else?

c) You have not got time to train or prepare others?
d) Overwork is part of your job,

3. What word would you most associate with delegation?

a) Risk?
b) Fear?
c) Guilt?
d) Trust?

4. If you did delegate tasks, would they be:

a) The most boring ones?
b) The least risky ones?
c) The most risky ones?
d) The ones that a subordinate could do just as well?

5. If you had to delegate an important job to a subordinate, would you:

a) Issue it as an order?
b) Be very apologetic?
c) Leave it to someone else to convey?
d) Present it as an opportunity?

6. When delegating to someone, do you:

a) Keep worrying that the job is not being done well?
b) Ask them to report back each time a decision is made?
c) Stipulate that, if anything goes wrong, it is your responsibility?
d) Tell them to come back to you only if there is a problem they cannot handle?

Effective delegation is about trusting your staff and colleagues and delegating authority – but not responsibility. If you answered 'd' to each question, you need read no further. If not, then have a look at my answers.

QUESTION 1: Delegation should never be forced on others, nor presented in a negative way. At best, it is an opportunity for career development. However much you delegate, the buck always stops with you.

QUESTION 2: If you do not trust your staff, you cannot truly delegate – you will always be involved in your judgements. It is all about learning to respect and trust your team so that tasks can be more evenly spread.

QUESTION 3: Delegation = trust.

QUESTION 4: Delegating the worst jobs is not worthy of the 'delegation'. You should delegate tasks that you would normally be able and prepared to do. It is not an excuse for offloading rotten jobs.

QUESTION 5: Good delegation needs to be presented as a positive benefit. How you 'sell' delegated tasks is most important. You should delegate the interesting and challenging jobs – and negotiate with the delegate.

QUESTION 6: Learn to let go. If you trust your subordinates, let them run with a task. If you feel any doubts about your capabilities, invest in training and staff development – it's cheaper than your suffering from stress-related illness.

7.1.9 Key action points to remember on delegation skills
- Invest time in **people**.
- **Anticipate**: strategy – competition – problems.
- **Think** about: 'tomorrow' – outstanding work – delegation.
- **Plan** your time well – four weeks ahead.
- Establish **start** times.
- **Delegate**: good for you and your team.
- **Meetings**: small agenda – small attendance – precise actions – punctual finish.
- Keep your secretary/clerical support **in the picture**.
- Ensure that your filing system makes for easy **retrieval**.
- Keep your desk for **work, not storage**

7.2 Appraisals

7.2.1 Performance appraisals

Appraisals are all too often the bane of a working life rather than something to look forward to and enjoy. Appraisals should be conducted once a year as an absolute minimum, with less formal, quarterly appraisals in the interim.

> **Networking hint**
> Appraisals should be a win–win experience: both parties should gain by it and feel a sense of satisfaction and achievement. Here is an opportunity both for managers and staff to assess each other's performance, build relationships and receive constructive feedback.

Benefits of the appraisal process accrue to both the individual, the line manager and the client. Let us look at each in turn.

7.2.2 Benefits to the individual
- Discussion of the job role in the context of job description.
- Assessing performance against agreed objectives.
- Opportunity to give and receive feedback.
- Having training needs identified.
- Opportunity to discuss career prospects and promotion.
- Future planning – understanding and agreeing objectives.
- Building relationship.
- Reinforcing the delegation process.
- On-the-spot coaching.
- Increasing motivation and improving morale.

7.2.3 Benefits to the line manager

- Evaluating performance (individual, team, organization).
- Making the best use of resources.
- Giving constructive feedback.
- Setting and clarifying objectives.
- Identification of training needs.
- Audit of team's strengths and weaknesses.
- Receiving feedback on management style.
- Exploring and resolving problems.
- Reducing staff turnover.

7.2.4 Benefits to the client

- Improved performance through commitment to the project.
- A minimum standard of good project management.
- Sharing of skills.
- Appropriate manpower utilization.
- Test of selection process.
- Reducing project staff turnover.
- Improvement to morale and motivation.

7.2.5 Preparation for the discussion

Preparation for the interview is essential if both parties are going to get the most out of it. You will need to think carefully about what you want to discuss, gather relevant information and focus on relevant issues. You will need to notify the person in writing and familiarise yourself with the individual's file and performance factors.

You will also need to think about the environment in which you conduct the appraisal. A neutral location is generally better than your office. Make sure that you have what you need in the room – water, tea, coffee – and ensure that you are both comfortable and that you will not be interrupted. And be sure to allow enough time.

7.2.6 Conducting the interview

During the interview:

- start on a positive note – emphasise what is working;
- use the 10:1 ratio for feedback – 10 positives to 1 negative;
- create a relaxed, positive atmosphere;
- review the purpose of the interview;
- use an agenda;
- encourage the role holder to talk;
- listen carefully;
- use open questions;
- keep to the agenda during the interview;
- ensure you cover all the key aspects of the role;
- discuss areas of improvement;

- avoid overcriticizing;
- deal with one topic at a time;
- summarise and maintain control throughout;
- discuss further training and career development needs;
- review and summarise main points, agree action plans;
- end on a positive note, thank each other for contributions.

7.2.7 Active listening

	Active	**Passive**
L	Look interested	Show no encouraging responses
I	Involve yourself by questioning	Ask irrelevant questions or assume
S	Stay on target	Become distracted or daydream
T	Test your understanding	Do not clarify or summarise
E	Evaluate the message	Do not connect/relate to other information
N	Neutralise your feelings	Have prejudices and make snap judgements

7.2.8 Constructive feedback

When giving constructive feedback remember to:

- balance praise and criticism – 10:1 ratio;
- be constructive;
- be factual and specific;
- seek clarification;
- maintain open communication;
- focus on behaviour, not personality;
- be prepared to give and receive;
- be honest;
- agree future changes/solutions.

Appraisals can be opportunities for change on the one hand or a damaging experience on the other. Correctly used, they can make the difference between a high-performing and motivated individual and one who does the bare minimum to get by.

7.2.9 Appraisee's Charter

I have the right to:

- receive my appraisal when it is due;
- be clear what is expected from me;
- have feedback on my performance;
- gather my own evidence;
- make genuine mistakes;
- contribute equally to agreeing objectives and standards;
- raise issues and concerns;
- consult others.

7.2.10 Appraiser's Charter

I have the right to:

- give feedback on performance;
- contribute equally to objectives and standards;
- consult others;
- say 'no' to unreasonable requests;
- adjourn the performance discussion;
- expect certain standards of work and behaviour.

7.3 Disciplining staff and problematic colleagues

If only your staff and colleagues worked harmoniously and efficiently all of the time! Wouldn't that be wonderful? What if they never complained, went sick or had 'attitude'? Can you imagine it? It must surely be every busy manager's dream. The reality, however, is somewhat different. People are fallible, staff and colleagues do make mistakes. Bullying, harassment and discrimination in the workplace do occur, even in the best of organizations. Unfortunately, it's usually the job of the manager to sort out the problem.

7.3.1 Setting a good example

Staff and colleagues are expected to have accountability when accepting a position on a project team. They should seek to maintain the standards of work and behaviour set out under the terms of the contract, and abide by their professional code of conduct. If they trip up once or twice, perhaps a gentle reminder is all that is required. If they fail to do what is expected of them on a regular basis, several adverse things happen. It costs the organization money, it upsets the balance of the project team ('If he gets away with it, why can't I?'), morale plummets and the project manager's headaches reach epic proportions.

> **Networking hint**
> Everyone at work is entitled to be treated with dignity and respect. Bullying, harassment and discrimination are in no one's interests and should not be tolerated anywhere in the workplace.

Bullying is usually characterised as offensive, intimidating, malicious or insulting behaviour. Harassment is generally unwanted conduct affecting the dignity of men and women, relating to age, sex, sexuality, race, disability, religion, nationality or any personal characteristic of the individual.

If prevention is better than cure, one good rule for project managers is to give staff and professional colleagues examples of what is regarded as unacceptable behaviour in the circumstances. Companies, whether large or small, should have policies and procedures for dealing with grievance and disciplinary matters. Staff should know to whom they can turn if they have a work-related problem. Project managers should be trained in all aspects of the organization's policies in this sensitive area.

7.3.2 Bullying and harassment

People who are bullied or harassed may seem to overreact to something fairly trivial. However, it could be the last straw following a series of incidents. The dangers of allowing such behaviour to go unchecked are that they create serious problems for the organization as a whole. These include poor morale, inharmonious staff or colleague relations, loss of respect for management, bad performance, low productivity, absences, resignations – all of which seriously damage the company's reputation.

When a member of staff makes frequent mistakes and exhibits inappropriate behaviour, and when their performance standards are falling way short of the company policy, this is not acceptable. They are showing contempt by not caring about their work, their company or the effect of their behaviour on their colleagues. Swift, decisive, corrective action needs to be taken.

'Why bother?' you may ask. Why risk ending up in front of an employment tribunal, with all the concurrent hassle and traumas? That is every project manager's nightmare. Why not take the ostrich approach and hope matters improve. Surely it's easier to leave things as they are. Well, actually, it isn't. No problem ever got smaller by leaving it alone. What starts as a minor dispute can develop into a full-blown crisis if ignored. By not confronting the issue, things only get worse.

7.3.3 Discipline versus punishment

Discipline should not be confused with punishment. Discipline is positive; punishing someone is to do with exacting a penalty. Effective discipline involves dealing with the shortcoming or misconduct before the problem escalates. Disciplining a staff member or colleague can be an informal or formal procedure, depending on the severity of the problem. If an informal approach is appropriate, counselling or training can provide a vital role in resolving complaints. Whichever the case, the important thing is to follow a fair procedure. When the issue involves a complaint about bullying, harassment or discrimination, there must be fairness to both the complainant and the person accused.

Set a good example. The faster you deal with the problem, the stronger example of management behaviour you are giving. If you allow time to elapse between the incident taking place and disciplining the staff member or colleague, the message you are sending is that you couldn't be bothered to do much about it. Maintain fair procedures for dealing promptly with complaints from staff and colleagues. Set standards of behaviour by means of an organizational statement to all staff or through the company handbook. Finally, let staff and colleagues know that complaints of bullying, harassment or discrimination will be dealt with fairly, confidentially and sensitively.

7.3.4 Possible solutions

Because of recent changes in the law, the prudent manager should consult the ACAS (Advisory, Conciliation and Arbitration Service) advisory handbook, *Discipline and Grievances at Work*. The ACAS code of practice, 'Disciplinary and Grievance Procedures', gives advice on good practice in disciplinary matters. This is something that is taken into account in cases appearing before employment tribunals.

First of all, identify the problem. Is the issue you are dealing with related to performance or misconduct? Try to deal with the matter as quickly as possible. Make sure you explain why the individual is being disciplined. Describe exactly what the unacceptable behaviour is. Here, it is essential to avoid generalities, so be as specific as possible. Detail what changes need to be made and outline the consequences if the unacceptable behaviour continues.

> **Networking hint**
> Stick to the facts; focus on the behaviour and not the person. Explain the effect their behaviour or actions are having on the rest of the company/department/unit.

In order to comply with the law, which requires fairness above all things, you will need to carry out a full investigation, giving the individual the opportunity to state their case. They are allowed to be accompanied to any interview or hearing by a colleague or company representative. Make sure you give an explanation for the disciplinary action and specify clearly the appeals procedure.

In the case where you suspect someone has made an unfounded allegation of bullying, harassment or discrimination for malicious reasons, it is essential that such allegations be investigated fully and dealt with fairly and objectively under the disciplinary procedure.

Depending on the outcome of the disciplinary procedure, reasonable action should be taken in relation to the facts. Penalties are not always necessary: sometimes it is more appropriate to offer counselling or training. In the severest cases, where bullying, harassment or discrimination amounts to gross misconduct, dismissal without notice may be the right course of action.

Where such issues arise, project managers should examine their company policies, procedures and working methods to see if they need to be improved. Useful contacts where further advice can be sought include ACAS, which you can find online at http://www.acas.org.uk, and the Equality Human Rights Commission, at http://www.equalityhumanrights.com.

7.3.5 Summary/check list

In conclusion, should you, as a project manager, have to discipline a member of your staff or report a professional colleague, remember the golden rule: keep a record of everything. You can't have too much documentation when taking personnel actions. Give fair warnings (along with notification of the consequences) and always in plenty of time. It is also important to give your staff or colleague enough time to respond, or rectify their behaviour. Make sure your company's policies are reasonable and that standards are achievable. Finally, make clear what avenues there are for appeal and that the staff member or colleague knows what they are.

The law in this area is complicated and all employers contemplating dismissal, or action short of dismissal, such as loss of seniority or pay, are required to follow a three-step statutory procedure. For more details see the website of the Department for Business, Enterprise and Regulatory Reform (BERR) at http://www.berr.gov.uk or consult your employment lawyer for specific advice.

7.4 Recruiting and selecting the right people

Another aspect to building excellent people skills is the ability to recruit staff. It is something every effective manager should be able to do. But finding and hiring the best applicant for a job is no easy task. With lots of people looking for work, particularly in the current economic downturn, it is challenging to have to pick the best person from a large number of candidates. Whether you are about to hire your first employee or have taken on staff many times before, you know the feeling – it is a leap in the dark. Recruitment and selection is a vital task that managers frequently have to fulfil. Get it right and everyone benefits. Get it wrong and the consequences are dire.

Anyone ambitious who wants their department or project team to grow sooner or later needs to take on staff. There comes a time when outsourcing has reached its limits, there is no time to finish the countless tasks – more hands are needed without doubt – and, to make the task seem less daunting (taking into account the growing complexity of the government's employment laws), you need to proceed with care. If you aren't completely confident about your ability to deal with people, this can be an unnerving prospect.

Let's take a step-by-step approach to the recruitment process. It should come in useful at some stage.

7.4.1 Defining the job to be done

Analyse the job and draft the job description. The creation of a clear and well-structured job description is an essential first step. Investing time at this stage is a good policy. Whether the position is a new one or you are filling an existing one, before starting the recruiting process be sure you know what standards you are going to use to measure your candidates.

Write down the description of the job, whether it is a newly created post or an existing position. What is the job title? What are the objectives and purpose of the job? What duties, responsibilities and tasks go with it? How does it fit with existing jobs? Where will it lead and what prospects can new staff be offered? Describe the reporting lines and working relationships. State the specific tasks, standards and responsibilities required. Detail the appraisal procedure and be clear as to the remuneration package and other benefits.

While a clear job description is fundamental to successful recruitment, the personal profile sets out the characteristics of the kind of person who might be qualified and suited to undertake the role. People are the core of any business. As you will know from experience, some people love their jobs while others do not exactly live for their work.

7.4.2 Specifying the profile of the likely candidate

Identifying the characteristics of the person who is most likely to be suitable for the position is useful. Descriptions such as 'hardworking', 'good attitude', 'experienced', 'stable', 'smart' and 'responsible' spring to mind. But how do you find such paragons?

7.4.2.1 Personal characteristics

This covers basic personal characteristics such as age, education, experience and specialist qualifications, for example fluency in a foreign language. The purpose here is to make the selection process manageable. Most employers wish to trawl a fairly wide area, but they are not keen to plough through hundreds of applications, most of which are unsuitable.

7.4.2.2 Character traits

Do you want someone creative, industrious, loyal or innovative? Aspects of character, such as these, are important attributes but much more difficult to measure accurately.

7.4.2.3 Motivational factors

Will the job suit someone who wants a steady routine or someone who wants something more challenging? What you need to look at here is what is likely to appeal to an applicant about the job. Is it suitable for someone who is ambitious, competitive, innovative and creative? These are set as a guide only.

7.4.2.4 Responsibility

Areas of responsibility relate to the aspects of character that make a candidate suitable for the post. Does the applicant have the ability to work on their own, care for others or give presentations to large audiences? Will they need to work as part of a team? Is 100 per cent accuracy essential in their work?

The worst-case scenario is to end up appointing someone who proves not able to do the job but is not so appallingly bad and therefore cannot be sacked. It is important to consider the kind of person you feel would be best suited to the position. Once you have given some time to these details, you can start the selection process.

7.4.3 Sources of candidates and methods of attracting the right person

From among the sources of attractive potential candidates, perhaps the most effective are to be found among the following.

7.4.3.1 Internal selection

This would be via the HR or Personnel Department – an internal advertisement of the position. Provided an effective training and selection programme is in place, it is possible to source and select for the new position from within the company. The advantage here is that the applicant is known to the manager, the applicant knows the company and has 'bought into' the culture of the firm. It is good for staff morale to see that inside promotion is possible. Only after you have exhausted your internal candidates should you look outside the company.

7.4.3.2 Referrals

If you are looking to fill a vacancy, make sure you let people know. Whether it is colleagues, friends, relatives or clients, many good candidates are sourced from referrals. Someone you know can give you great insights into the applicant's strengths and weaknesses and character. You will get far more information than you would from résumés alone.

7.4.3.3 External advertising

If you are writing a job advertisement, make sure the copy describes the actual job to be done, and describes the organization in terms of what it does and its style and culture. It also needs to state clearly a specific salary range and the nature and qualifications of the candidates sought. Situations-vacant advertisements are relatively inexpensive and can get your job publicized over a wide area. This may have its advantages but the disadvantage is that you may have to sort through literally hundreds of applications to find a few good ones to shortlist.

7.4.3.4 Temp agencies

If in doubt, hire a temp or locum. This will give you some relief if the work is piling up while the recruitment process is under way. It also provides the opportunity to try out employees before you hire them. If you like your temp, ask the agency if you can recruit them for a nominal fee or after a certain period of time.

7.4.3.5 Recruitment consultants

These can be used for sourcing applicants for a specialized position or if you simply do not have the time to go through the whole process of recruitment and selection yourself. The agency carries out the advertising, recruiting and screening of applicants, providing you with a shortlist of perhaps five people to interview.

7.4.3.6 Executive selection and headhunters

The higher the level of the position you are seeking to fill, the more appropriate it may be to seek assistance from one of the executive search companies or headhunters. They have great experience and are able to select candidates to the highest standard.

7.4.3.7 The Internet

Many people regard the Internet as an effective place to advertise vacancies. Web pages allow you to present large amounts of information on your company and your job opportunities. The Internet is always accessible and reaches a global audience.

7.4.4 Assessing written personal details

When considering applicants' CVs bear in mind they can be quite distinctive.

- A **targeted CV** draws attention to the applicant's skills and focuses on the qualities that make her the right person for the particular position.
- A **chronological CV** is one that summarises the qualifications and career experience of the candidate. This form of CV is popular with local authorities, central government and more traditional employers.

- An **experience-based CV** is valid for individuals who work in specialized areas of employment. It describes their track record to date.

CVs should be precise and list a candidate's achievements with detailed points. Beware CVs that generalize – this could indicate a weak candidate. There is no need for large amounts of personal detail to be added to a CV. It should be a piece of personal marketing literature, focusing on the product – in other words, the skill the candidate offers.

CVs that are gimmicky are risky. A CV should look attractive, clean and professional. Trendy typefaces and coloured ink are not appropriate. If a CV contains spelling mistakes, it could indicate a careless individual.

Be vigilant at checking accuracy of CVs – employment history, qualifications and skills.

7.4.5 Systematic approach to interview

When it comes to interviewing candidates, you must ask the right questions. Hiring the right people is essential to the growth and success of your business. You, as manager, need to get your interview technique right. This means asking loaded questions that will reveal the information you need to make an informed decision.

At the outset, welcome the applicant, then begin by summarizing the position. Ask the prepared questions and use the candidate's answers to evaluate strengths and weaknesses. Conclude the interview after allowing the candidate the opportunity to ask any questions they wish. Advise them of when you will be making your selection.

Take time to prepare your questions and make notes of the applicant's answers.

'Why are you here?': This may elicit the answer, 'Because I want a job with your firm.' But it could give you other information that you would never have gleaned had you not asked in the first place.

'What can you do for us?': An important question and one that candidates should be prepared for. It is quite common for applicants today to approach the recruitment process on the basis of, 'What can your company do for me?' and this question redresses the balance.

'What kind of person are you?': You need to know – after all, if they're recruited, you will be spending considerable periods of time in their company. You want to employ someone who will be congenial most of the time he is at work.

'If you stayed with your current company, what would be your next move?': This is a question designed to extract information on several levels. It should reveal a sense of what the applicant expects but also why they want to move on. If, for example, they say they want to be a manager but the person above them has been there for 25 years, you can move on with the interview. If they were to say they hoped to be promoted within six months, why are they leaving that job? You should try to elicit the real reason why they want to leave that company.

'What do you consider makes you exceptional compared with others?': A difficult question for most people, because applicants have a tendency to be uncomfortable praising themselves. If the answer is given in a reasoned manner, the applicant may have a good degree of self-esteem and some courage. A timid response could indicate a reticent type – not one fit for a challenging role within your company. Beware the applicant who launches into a lengthy monologue about why the world revolves around them. The overactive ego could spell disaster if this applicant is aiming to fill a position where teamwork is a requisite.

'Describe your greatest achievement to date': If the applicant can recall quickly and with detail a satisfying and recently accomplished project, which they recount in a measured, comprehensive way, you may have a winner here. Any applicant who is quick enough to think on their feet and produce the anecdote without hesitation is likely to be an asset to your company.

'Do you need many hours a week to get your work done?': This question is designed to elicit the work ethic of the potential employee. If they expect to put in long hours with your company, this could indicate that they're staying late to do extra work, or merely that they work inefficiently. A discussion as to working habits can reveal how they will fit in with the rest of the employees. A company where it is normal to stay until 7 p.m. would not suit someone with a strict eight-to-five mentality.

'What sort of salary are you expecting?': There is no point in having gone through the selection process only to reach the end of the interview and find out that your idea of a competitive salary and benefits package is so far removed from the candidate's that you seem to be on different planets. You may not be able to offer enough in terms of salary, but try putting together a generous benefits package – including pleasant office, corporate membership of a health club, impressive job title, generous holidays, health insurance, pension.

> **Networking hint**
> Don't forget to keep copious notes as you interview each of the candidates. It will be impossible to remember who said what and your written notes will be an essential aid when evaluating the applicants.

There are one or two topics on which it is not appropriate to question applicants: these relate to the applicant's race, skin colour or national origin; also anything to do with marital status, religion and criminal record. Personal details such as height, weight, financial status and disabilities are also not necessary. These questions do not relate to how the applicants perform their jobs and are best avoided.

7.4.6 Assessment and checks

Check, check and check again. References, skills, previous employment history. Do the dates on the CV match up with other facts? Was that extended holiday actually spent at Her Majesty's pleasure?

It is quite surprising how many people exaggerate their education experience. This is a good place to start your checks. If it is found that the application is inaccurate about one thing, it is likely that the rest of the CV is too.

Don't be hesitant about calling previous employers for information about an applicant. You should get more detail from a supervisor or manager than from the HR or

Personnel Department, which is more likely simply to confirm the dates the applicant was employed. With regard to skills, if the job requires a high proficiency in presentation abilities, ask the applicant to demonstrate their expertise. If they need to write well, look at some examples. If the requirement is for fluent French, set a test for the applicant.

7.4.7 Final selection and appointment

When reviewing the information about all the candidates, you will be glad to look at the notes you made – you did take notes, didn't you? How does each applicant stand up against your original criteria for the position? Is there an outright winner? How many losers? Sort them into categories: winners, possible winners and losers. Be objective. Don't be influenced by irrelevant elements such as clothes or hairstyles.

If there is more than one candidate with equally good qualifications, it may be necessary to go for a second or third round of interviews. This may seem time-consuming or expensive, but it is better than making a wrong appointment and living to regret it. If you really can't be sure, go with your gut instincts. Although they may seem equally matched in skills and experience, you will probably have a feeling that one is more suitable than the other. If so, allow your intuition free rein.

It is sometimes wise to hire people for their personalities rather than for their skills and qualifications. Whereas it is difficult to change someone's personality, it is not impossible to teach them new skills or train them in certain techniques. If in doubt, don't be afraid to ask advice. Managers hiring staff must be certain of the decision they are about to make. Ask your mentor, or an experienced director of a well-established company. Departmental managers cannot afford to make mistakes and most people are happy to help if requested to do so.

Never hire someone on the basis that 'he's the best of a bad bunch'. This is potentially disastrous. Better to repeat the whole process, because rarely do people make a miraculous change for the better once they are appointed. Once you have made your decision, telephone the successful applicant as soon as possible and offer your first choice the job and secure his acceptance. If he is no longer available, go to candidate B. Hopefully, you will be able to hire someone from among your selection of 'winners'.

Don't forget to communicate with the unsuccessful candidates. It is old-fashioned courtesy, but it does cost nothing to be polite and it will make them feel a lot better. It could be a short letter saying you will consider them for alternatives (if they might suit another position in the company) or will keep their CV on file (and telling them you are doing so).

Finally, don't forget that, if your department or team grows the way it should, your new employee could be a manager one day. They will be running the office in your absence. By looking for someone who possesses that extra spark, someone who would be happy to take responsibility and act on their own initiative, you could be choosing your successor.

Chapter 8
Is it working OK?

Being part of a team, managing others, motivating staff and keeping everyone in line – this chapter covers the remaining aspects of working relationships within the company. What you probably want to do most (if you're managing other people) is to win over the hearts and minds of the people with whom you work. This is not just for survival – although it would make life a lot easier – but it's because you are keen to succeed and get one of the steepest parts of the learning curve over and done with.

Becoming a recognized team player is something you must and should be able to do if you are to have effective workplace relationships. You will not achieve this by being aloof. If you want to go (and be carried) along with the crowd, the best you can do is:

- show leadership qualities;
- get involved;
- not shy away from getting your hands dirty (often);
- know what is going on so that you are able to do all this.

People support those they feel understand their situation, in other words those who build rapport with them. If you can show that you have experience of being a team player, most people will readily accept you. It is then up to you to make sure you can genuinely play your part and become a willing and active participant.

One way of achieving a quick win here is when a crisis strikes. People will like it if you rise to the occasion in an emergency. In an 'all hands to the pumps' situation, don't hang back. Get stuck in and don't pick the easiest (cleanest) task. There is nothing more helpful towards gaining respect from other team members than willingly taking a turn at the less exotic tasks.

8.1 Meeting the team

If you have to form a team, or join an existing one, treat it like another interview. Be as professional as you can and ask questions. The team will be glad to tell you what you need to know. Meeting the team isn't just a matter of quickly sizing people up: it can be very useful. The key to meeting the team is to be relaxed. Most of them will be your peers and you will be under scrutiny just as much as you are checking them all out. The team could ask you all sorts of questions – but do make the most of the opportunity to be equally inquisitive and observant.

If you ask them about their work, you will see how they interact with each other. If you can work out how the hierarchy operates it will be a help. For instance, is everyone equal? Or is there very much a 'top dog'? If you are joining a big team, they will be thinking about how your skills can bring an overall improvement to the set-up. People always want to know how your presence can help them. Some team members may have different expectations or needs.

This meeting is a two-way process. You want to make a strong impression but trying too hard and not being yourself is not a good idea. Adapt your behaviour, by all means. After all, this process is about matching. If after the initial meeting you have a few worries, ask your manager how she thinks you will fit in with the team. If she has had any feedback from them she should use this opportunity to share it with you.

8.2 Working together

You may be new to the experience of working as part of a team, in which case it will help if you make this clear early on. If the rest of the group know this, they will give you some support and help. Becoming part of a team means becoming involved with everyone in a positive way. Interacting with the rest of the team and being a 'player' ensures that together you all succeed.

> **Networking hint**
> The most important thing when working as part of a team is to be honest with yourself and in turn with others.

The organizational structure of a team is important. Who does what, how one job relates to another, the lines of reportings and communication – all of these affect effectiveness. This is something to assess early on and watch on a regular basis. Whether you are a team player or a team leader, here are a few points about how teams should work.

- The team structure should fit the tasks to be done.
- Any changes should be made on a considered basis.
- Those responsible for the changes should express them in a positive way, otherwise they may be viewed with suspicion.
- The team leader should keep the organization under review to ensure a good 'fit' between the team and its task, so that it continues to perform well (external as well as internal changes or pressures can affect this).
- Teams sometimes need a bit of fine-tuning – keeping an eye on tasks, individuals and the team as a whole.

Any, even slight, incongruities about the way people are organized can easily dilute overall effectiveness. Team leaders should not make change for change's sake, but neither should you expect things to remain without needing change for ever.

8.3 Team dynamics

A quick and simple way if you're experiencing team work for the first time is to remember the dynamic process of working on a project with a group. Think of the following five words: *forming, norming, storming, performing, mourning*.

- **Forming**: The group have to get together and agree the way forward. Often there is a uniting cause or goal that binds the team that starts the process of cohesion.
- **Norming**: When the group have agreed the ways forward and resolved the structure of working together, they begin a working relationship.
- **Storming**: As the group settle down to the task, differing opinions, styles and ways of working become apparent and friction appears. This needs to be addressed.
- **Performing**: As this working relationship matures, and everyone in the team has trust and faith in each other and the team as a whole, they perform at their best.
- **Mourning**: This fifth step is often forgotten, but, when a process of change comes to an end, the goal has been reached, there is a sense of relief. But in some cases it also creates a sense of anticlimax and loss. For example, if the change has meant redundancies or relocation, there is a physical process of ending rather than the celebration of a job well done.

If you have recently joined a team, where are you in the various stages of team development, keep this exercise in mind to review from time to time. If a new person enters the team, or other factors relating to the project suddenly appear, you may find that the group has to re-form, norm and storm all over again to get the performing to its best level.

8.4 Self-sufficiency and group dynamics

If you organize things so that people are suitably self-sufficient, it saves time and promotes goodwill. Remember that having responsibility is motivational – and that means that people tend to do best those things for which they see themselves as having personal responsibility.

This means thinking in terms of – and using – two distinct levels of self-sufficiency in how people work: involvement and empowerment.

8.4.1 Involvement

First, it is necessary to create involvement in ways such as consultation, giving good information, and making clear that suggestions are welcome and that experiment and change in how people do things is a good thing.

This provides the opportunity to contribute beyond the base job.

8.4.2 Empowerment

This goes beyond simple involvement. Empowerment adds the authority to be self-sufficient (making your own decisions) and creates the basis for people to become self-sufficient on an ongoing basis. In a sense, empowerment creates a culture of involvement and gives it momentum. We look in more detail at empowerment later in the chapter.

8.5 The power of responsibility

Together, involvement and empowerment create an environment in which people can have responsibility for their own actions.

> **Networking hint**
> Responsibility cannot be given: it can only be taken. Thus only opportunity to take it can be given.

Creating a situation in which people take responsibility for their work demands:
- clear objectives (people knowing exactly what they must do and why);
- good communications;
- motivation (to show the desirability of taking responsibility for the individual as well as the organization);
- trust (having created such a situation, as team leader, you have to let people get on with things).

A team's enjoyment of involvement in what they do, and their having the authority to make decisions and get the job done is the best recipe for successful management and a healthy and contented workforce.

8.6 Strengthening the team

To recap for a moment, a successful team is one that:
- is set up right;
- responds to the responsibility it has for the task;
- seeks constant improvement (and does not ever get stuck 'on the tramlines');
- sees its manager as a fundamental support to its success.

A team situation will do well and is more likely to go on doing well than a group of co-workers just 'told what to do'. You, the team leader, have a role as catalyst – constantly helping the team to keep up with events, to change in the light of events and succeed because they are always configured for success.

8.7 Motivation

8.7.1 A brief overview

For any effective manager, it helps to know something of how motivation works. The key is affecting the 'motivational climate' by taking action to reduce negative influences while increasing the positive ones.

Networking hint
The state of motivation of a group or individual can be likened to a balance. There are pluses on one side and minuses on the other. All vary in size. The net effect of all the influences at a particular time decides the state of the balance and whether – overall – things are seen as positive, or not.

Changing the balance is thus a matter of detail with, for example, several small positive factors being able to outweigh what is seen as a major dissatisfier.

8.7.1.1 Reducing negative influences
Views about many factors potentially dilute the good feelings people have about their jobs. These include company policy and administrative processes, supervision (that's you, unless you are careful!), working conditions, salary, relationships with peers (and others), impact on personal life, status, security. Action is necessary in all these areas to counteract any negative elements.

8.7.1.2 Increasing positive influences
Here, specific inputs can act to add strength to positive feelings. These can be categorized under the following headings: achievement, recognition, the work itself, responsibility, advancement and growth.

Many things contribute, from ensuring that a system is as sensible and convenient to people as possible (reducing negative policy/working conditions), to just saying 'Well done' sufficiently often (recognizing achievement).

8.7.2 Motivating the team
Don't underestimate your staff. Their views can enhance everything – methods, standards, processes and overall effectiveness.

'United we stand, divided we fall' is an expression which comes to mind when considering the structure of teams. If you can involve people on broad issues, you will find this is motivational for all concerned.

Resolve now – right now – that you will make motivation a priority. Motivation makes a difference – a big difference. People perform better when they feel positive about their job. You should:

- recognize that active motivation is necessary;
- resolve to spend regular time on it;
- not chase after magic formulae that will make it easy (there are none);
- give attention to the detail;
- remember that you succeed by creating an impact that is cumulative in effect and tailored to your people.

Your intention should be to make people feel, individually and as a group, that they are special. Doing so is the first step to making sure that what they do is special. As team leader, or manager, you are not paid to have all the ideas necessary to keep your section working. However, you are paid to make sure that there are enough ideas to make things work and continue working. You should use your people and make it clear to them that you want and value their contributions.

8.7.3 Obtaining significant motivational results

Team leaders need to make it clear from the beginning that they are concerned that people get job satisfaction. Major schemes can wait. Early on, do these things.

- **Take the motivational temperature**: Investigate how people feel now, this is what you have to work on.
- **Consider the motivational implications of everything you do**: Put in a new system, make a change, set up a new regular meeting. Reflect: What will people think about it? Will they see it as positive?
- **Use small things – often**: For example, if asked if you have said 'Well done' often enough lately, you must always be able to answer 'Yes' honestly.
- **Never be censorious**: You must not judge other people's motivation by your own feelings. Maybe they worry about things that strike you as silly or unnecessary. So be it. The job is to deal with it, not to rule it out as insignificant.

> **Networking hint**
> Create the habit of making motivation a key part of your management style and doing so will stand you in good stead. If you care about the people, it will show.

8.7.4 Getting the best from the team

Recognize from the beginning that your effectiveness depends on your team and on the interaction of three separate factors. If you are team leader, you must:

- ensure continuous task achievement;
- meet the needs of the group;
- meet the needs of individual group members.

This balance must always be kept in mind (though some compromise may be necessary). Your own best contribution to getting things done is ideally approached systematically.

- Be clear exactly what the tasks are.
- Understand how they relate to the objectives of the organization (in the short and long term).
- Plan how they can be accomplished.
- Define and provide the resources needed for accomplishment.
- Create a structure and organization of people that facilitates effective action.
- Control progress as necessary during task completion.
- Evaluate results, compare with objectives and fine-tune action and method for the future.

There are three checklists here that can be used in conjunction with what has been set out above. These apply not only to team leaders, but should also be useful in helping you to integrate into the group.

8.7.4.1 Checklist 1: Achieving the task

Ask yourself the following.

- Am I clear about my own responsibilities and authority?
- Am I clear about the department's agreed objectives?
- Have I a plan to achieve these objectives?
- Are jobs best structured to achieve what is required?
- Are working conditions/resources suited?
- Does everyone know their agreed targets/standards?
- Are the group competencies as they should be?
- Are we focused on priorities?
- Is any personal involvement I have well organized?
- Do I have the information necessary to monitor progress?
- Is management continuity assured (in my absence)?
- Am I seeing sufficiently far ahead and seeing the broad picture?
- Do I set a suitable example?

8.7.4.2 Checklist 2: Meeting individual needs

Ask yourself if all members of the team:

- feel a sense of personal achievement from what they do and the contribution it makes;
- feel their job is challenging, demands the best of them and matches their capabilities;
- receive suitable recognition for what they do;
- have control of areas of work for which they are accountable;
- feel that they are advancing in terms of experience and ability.

Many questions stem from this about:

- what people do;
- how they do it;
- how what they do is organized;
- how they feel about it.

It is worth thinking through what you need to ask in terms of your own particular team and any new players who may recently have joined.

8.7.4.3 Checklist 3: Team maintenance

To involve the whole team pulling together towards individual and joint objectives, ask 'Do I, as team leader':

- set team objectives clearly and make sure they are understood?
- ensure standards are understood (and that the consequences of not meeting them are understood and approved)?

- find opportunities to create team working?
- minimize any dissatisfactions?
- seek and welcome new ideas?
- consult appropriately and sufficiently often?
- keep people fully informed (about the long and the short term)?
- reflect the team's views in dealing with senior management?
- accurately reflect organizational policy to the team and in their objectives?

An analytical approach to these areas is the foundation to making the team operation work effectively – and thus to getting tasks done effectively.

8.7.5 Consistency counts

People work successfully with all types of people – the tough and the tender. However, nothing throws them more than a team member who runs hot and cold. This problem is at its worst when the team leader shows these traits. Sweetness and light one minute (and being ready to listen and consult) and doom and gloom the next (demanding that people 'do as I say'). Early on you may need to experiment a little with how you deal with things. That apart, you should try to adopt a consistent style. For example, let people know that:

- you will always make time for them (soon, at an agreed time, if not instantly);
- you never prevaricate (decisions may not be made instantly – if they need thought or consultation – but nor will they be endlessly avoided; if there must be some delay, tell people why and when things will be settled);
- you will make sure they understand how you approach things and what your attitude is to problems, opportunities and so on; while solutions may, doubtless should, be different, your method and style of going about things should be largely a known quantity.

People like to know where they are, and work better when they do.

8.7.6 Why motivation matters

There is considerable research to show that people who are, to put it simply, happy in their work will perform better than those who are not. Many busy managers have, as part of their job, the need to get results *through* other people, rather than *from* them. If this is so in your case, the motivational state is important. In terms of both productivity and quality of action, maximizing motivational feeling will assist performance. If you are building strong relationships in the workplace, you will find that being able to motivate people is a huge asset.

It should be remembered that it is easy for any dilution of motivation to act to reduce performance – something that ultimately reflects on a team manager. Multiply the effects, either positive or negative, by the number of people reporting to you and you see the real importance.

If you are a team manager you must act not just to ensure that people perform well on their particular project, but that they do so consistently and reliably.

Good motivation also acts to make sure that people are as self-sufficient as possible, able to make decisions – good decisions – on their own and take action to keep things running smoothly. If you have to check every tiny detail and issue moment-by-moment instructions, neither productivity nor the quality achieved is likely to be as good as it might be. There is every difference in the world between people being *able* to do something and do it well, and being *willing* to do it and do it well.

> **Networking hint**
> Managers need to motivate people, rather than leave them to their own devices. Motivation, like so much else in people management, does not just happen. It must be recognized as an active process, one that you need to allow some time for on a continuing basis.

8.7.7 The fundamentals of motivation

Now we have looked at some of the how-to, let's look at *why* these things are important, by examining how motivating came about. The theory of motivation is extensive and this is not the place to do otherwise than recap some essential principles. Many people, certainly in years gone by, took the view that getting performance from staff was a straightforward process. You told them what to do, and they did it. If that was, for some reason, insufficient, it was backed by the power of management – effectively by coercion.

Management by fear still exists. In any economy with less than full employment – particularly relevant in the current climate – the ultimate threat is being out of a job. But, whether the threat is subtle or specific, whether it is just an exaggerated form of arm twisting or out-and-out bullying, even if it works (at least in the short term), it is resented. Your job as a manager is not simply to get things done: it is to get things done *willingly*. The resentment factor is considerable. People fight against anything they consider to be an unreasonable demand – so much so that the fighting may tie up a fair amount of time and effort, with performance ending up as only the minimum people 'think they can get away with'.

If people *want* to do things and are encouraged to do things well, only then can they be relied on actually to do them really well. Motivation provides reasons for people to want to deliver good performance. If this sounds no more than common sense, that is because it is. For example, are you more likely to read on if you're told that, if you do not, someone will come round to your house and break all your windows? Or, if you're persuaded that you will find doing so really useful and offer you some sort of tangible reward? (It is intended that you will find it useful, incidentally, but, sadly, there is no free holiday on offer.) Motivation works because it reflects something about human nature, and understanding the various theories about this is a useful prerequisite to deploying motivational techniques and influencing staff behaviour.

8.7.8 Theory X and Theory Y

The first of the classic motivational theories worthy of some note was documented by Douglas McGregor. He defined the human behaviour relevant to organizational life as follows.

- **Theory X**: This makes the assumption that people are lazy, uninterested in work or responsibility and thus must be pushed and cajoled to get anything done in a disciplined way, with reward assisting the process to some degree.
- **Theory Y**: This theory takes the opposite view. It assumes people want to work. They enjoy achievement, gain satisfaction from responsibility and are naturally inclined to seek ways of making work a positive experience.

There is truth in both pictures. What McGregor was doing was describing extreme positions. Of course, there are jobs that are inherently boring and mundane, and others that are obviously more interesting, and it is no surprise that it is easier to motivate those doing the latter. That said, however, it is really a matter of perspective. There is an old, and apocryphal, story of a despondent group of convicts breaking rocks being asked about their feelings concerning the backbreaking work. All expressed negative feelings, except one, who said simply, 'It makes it bearable if I keep the end result in mind: I'm helping to build a cathedral.'

Whether you favour Theory X or Y – and Theory Y is surely more attractive – it is suggested that motivation creates a process that draws the best from any situation. Some motivation can help move people from a Theory X situation to a Theory Y one, so it is easier to build on positive Theory Y principles to achieve still better motivational feeling and still better performance – and your communication should reflect this fact.

8.7.9 Herzberg's motivator/hygiene factors

This theory leads to a view of the process that links much more directly to an action-based approach to creating positive motivation. The American psychologist and motivation specialist Frederick Irving Herzberg (1923–2000) described two categories of factor: first, the hygiene factors (those dissatisfiers that switch people off if they cause difficulty); and, second, the motivators (factors that can make people feel good).

Let's consider these in turn.

8.7.9.1 Dissatisfiers (or hygiene factors)

These he listed, in order of their impact, as follows:

- company policy and administrative processes;
- supervision;
- working conditions;
- salary;
- relationship with peers;
- personal life (and the impact of work on it);

- status;
- security.

All are external factors that affect the individual (because of this they are sometimes referred to as *environmental* factors). When things are without problem in these areas, all is well motivationally; if there are problems, they all contain considerable potential for diluting any positive motivational feeling.

It should be noted here, in case perhaps it surprises you, that salary is part of this list. It is a potential dissatisfier. Would you fail to raise your hand in answer to the question: would you like to earn more money? Most people would certainly say yes. At a particular moment an existing salary may be acceptable (or unacceptable), but it is unlikely to turn you on and be a noticeable part of your motivation. So too for those who work for you.

It is, for instance, things in these areas that give rise to gripes and to a feeling of dissatisfaction that rumbles on. If the firm's parking scheme fails to work and you always find someone else in your place – perhaps someone more senior whom it is difficult to dislodge – it rankles and the feeling is always with you.

There are many things that spring from these areas for managers to work at, and getting them right can provide a positive boost to the motivational climate. The restriction here is that these things are not those that can add powerfully to positive motivational feeling. Get things right here and demotivation is avoided. To add more you have to turn to Herzberg's second list.

8.7.9.2 Satisfiers (or motivators)

These define the key factors that create positive motivation. They are, in order of relative power:

- achievement;
- recognition;
- the work itself;
- responsibility;
- advancement;
- growth.

It is all these factors, whether positive or negative and stemming from the intrinsic qualities of human nature, that offer the best chance of being used by management to play their part in ensuring that people want to perform, and perform well. For example, put in a new system – let's say asking people to fill in a new form on a regular basis – and, if it is not made clear why it is useful, people will be demotivated (because it relates to the list of dissatisfiers – specifically policy and administration – above).

> **Networking hint**
> Communication is a vital part of this picture. Every piece of communication can have motivational overtones – and probably will.

Similarly, a wealth of different communications all affect the motivational climate, jogging the overall measure of it one way or the other. Here are some examples.

- **Job descriptions, clear guidelines and adequate training** all give a feeling of security, without which motivation suffers.
- **Incentives** will work less effectively if their details are not clearly communicated. For instance, an incentive-payment scheme may be allowed to seem so complicated that no one works out how they are doing and motivation suffers as a result.
- **Routine jobs** can be made more palatable by communicating to people what an important contribution they make.
- **Job titles** may sensibly be chosen with an eye on how they affect people's feelings of status as well as acting as a description of function. 'Sales executive' may be fine and clear to customers, but most staff prefer titles such as 'account service manager').

Furthermore, the same essential act can be changed radically in terms of the effect it has motivationally just by varying the way in which communication occurs. For example, the simplest and least expensive positive motivational act you, as manager, can engage in is probably uttering the simple phrase 'Well done!' – and which of you can put a hand on your heart and say you do even *that* sufficiently often?

Consider some different ways of doing it, listed in what is probably an ascending order of motivational power:

- saying 'Well done!' one to one;
- saying 'Well done!' in public, in an open-plan office, say;
- saving it to be said at an 'occasion' (be it a departmental meeting or a group taking a coffee break together);
- saying it (in one of the ways listed above) and then confirming it in writing;
- getting the initial statement (however it may be done) endorsed by someone senior;
- publishing it (say in a company newsletter).

The implications here are clear. Not only is motivation itself primarily executed through communication, but the precise form of that communication needs to be borne in mind and contributes directly to the effect achieved.

8.7.10 Producing positive results

It may seem from what has been said already that motivation is a complex business. To some extent this is so. Certainly it is a process affected by many, and disparate, factors. The list of factors affecting motivation, for good or ill, may be long, and that is where any complexity lies, but the process of linking to them in terms of action is often straightforward.

The very nature of people and how their motivation can be influenced suggests five important principles for the manager dedicated to actively motivating people. These are as follows.

8.7.10.1 Principle 1: There is no magic formula

No one thing, least of all money, provides an easy option to creating positive motivation at a stroke, and anything that suggests itself as such a panacea should be viewed with suspicion.

8.7.10.2 Principle 2: Success is in the details

Good motivation comes from minimizing the factors that tend to create dissatisfaction, and maximizing the effect of those factors that can create positive motivation. *All of them* in both cases must be considered; it is a process of leaving no stone unturned, with all those found able to contribute to the overall picture being useful to utilize.

At the end of the day, what is described as the motivational climate of an organization, department or office is the sum of all the pluses and minuses in terms of how individual factors weigh in the balance, and communication plays a key role.

8.7.10.3 Principle 3: Continuity

The analogy of climate is a good one. As a small-scale example of this, bear in mind a glasshouse. Many factors contribute to the temperature inside. The heating, windows, window blinds, whether a door or window is open, whether heating is switched on and so on. But some such things – whatever they are – are in place and contributing to the prevailing temperature *all the time*.

So too with motivation. Managers must accept that creating and maintaining a good motivational climate takes some time and is a continuous task. Anything, perhaps everything, they do can have motivational side effects. For example, as was mentioned, a change of policy may involve a new system and its use may have desirable effects (saving money, say), but, if complying with the system is seen as bureaucratic and time-consuming, the motivational effect may be negative, despite results being changed for the better.

Overall, the trick is to spend the minimum amount of time in such a way that it secures the maximum positive effect.

8.7.10.4 Principle 4: Time-scale

Another thing that must be recognized is the differing time-scales involved here. On the one hand, signs of low motivation can be a good early warning of performance in peril. If you keep your ear to the ground, you may be able to prevent negative shifts in performance or productivity by letting signs of demotivation alert you to the coming problem. The level of motivation falls first, performance follows.

Similarly, watch the signs after you have taken action aimed at actively affecting motivation positively. Performance may take a moment to start to change for the better, but you may well be able to identify that this is likely through the signs of motivation improving. Overreacting because things do not change instantly may do more harm than good.

If motivation is improving, performance improvement is usually not far behind.

So, the timing of communication is vital, too. A busy moment and something allowed to go by default may lead to problems at some point in the future.

8.7.10.5 Principle 5: Bear others in mind

There is a major danger in taking a censorious view of any motivational factor – positive or negative. Most managers find that some, at least, of the things that worry their staff, or switch them on, are not things that would affect *them*. No matter. It is the other people who are important. If you regularly find things that you are inclined to dismiss as not of any significance, be careful. What matters to you is *not* the same as what matters to others.

If you discover something that can act for you influencing your people, however weird or trivial it may seem, use it. Dismissing it out of hand – and, in communications terms, say, failing to explain something adequately – just because it is not something that you feel is important will simply remove one factor that might help influence the motivational climate. It will make achieving what you want just a little more difficult. At worst, it will also result in your being seen as uncaring. Similarly, what is important to you may not be to others. This is an important factor that any manager forgets at their peril.

A further aspect of motivation now needs to be added: that concerned specifically with involving people.

8.7.11 Involving people

The word 'empowerment' enjoyed a brief vogue in the mid-nineties, as one of a succession of management fads that, if you believe the hype, solve all problems and guarantee to put any organization on the road to success. If only! On the other hand, there is sense in the idea of involvement that is essentially the meaning of empowerment. It may not solve everything, but it is useful and it does provide additional bite to the prevailing motivational feeling.

8.7.11.1 Behind empowerment

Empowerment does not allow managers to abdicate their responsibility, nor does it represent anarchy, a free-for-all, where anything goes.

But, for example when handling a customer's problems, consider what must lie in the background; staff must be able to:

- be proficient at handling complaints so as to deal with anything that might occur promptly, politely and efficiently;
- have in mind typical solutions and be able to improvise to produce better or more appropriate solutions to match the customer situation;
- know the system: what cost limit exists, what documentation needs completing afterwards, who needs to be communicated with, etc.

The systems – rules – aspect is, however, minimal. There should be no need for forms to be filled in beforehand, no hierarchy of supervisors to be checked with. Most of what must happen is left to the discretion of the individual members of staff.

The essence of empowerment is a combination of self-sufficiency based on a solid foundation of training and management practices that ensure that staff will be able to do the right thing.

8.7.11.2 Letting go

Often, on training courses, the room is full of managers tied, as if by umbilical cords, to their mobile telephones or pagers. Many of the calls that are made in the breaks are not responses to messages: they are just to 'see that everything is all right'. Are such calls, or the vast majority of them, really necessary?

The opposite of this situation is more instructive. See if this rings a bell. You get back to the office after a gap (a business trip, holiday, whatever). Everything seems to be in order. When you examine some of the things that have been done you find that your view is that staff have made exactly the right decisions. And yet you know that, if you had been in the office, *they would have asked you about some of the issues involved*. Some of the time staff empower themselves, and, when they do, what they do is very often right.

All empowerment does is put this kind of process on a formal footing. It creates more self-reliant staff, able to consider what to do, make appropriate decisions and execute the necessary action successfully.

Perhaps you should all allow this to happen more often and more easily.

8.7.11.3 Making empowerment possible

Empowerment cannot be seen as an isolated process. It is difficult to view it other than as an integral part of the overall management process.

You can set out to create a feeling of empowerment only by utilizing a range of other specific management processes to that end. The process perhaps starts with attitude and communication. What degree of autonomy do your staff feel you allow them? If they feel restricted and, at worst, under control every moment of the day, they will tend to perform less well. Allowing such feeling is certainly a good way to stifle initiative and creativity.

So you need to let it be known that you expect a high degree of self-sufficiency, and manage in a way that makes it possible. All sorts of things contribute, but the following – all aspects of communication – are certainly key.

Clear policy: Empowerment will work only if everyone understands the intentions of the organization (or department) and their role (clear job descriptions), so as to allow them to put any action they may need to decide upon in context.

The other requirement of an empowered group is an absence of detailed rules to be followed slavishly, but clear guidelines about the results to be targeted.

Clarity of communication: This has been mentioned before, but is especially important in the context of motivation. Any organization can easily be stifled by lack of, or lack of clarity in, communication; an empowered group is doubly affected by this failing.

Little interference: Management must set things up so that people can be self-sufficient, and then keep largely clear. Developing the habit of taking the initiative is quickly stifled if staff know nothing they do will be able to be completed without endless checks (mostly, they will feel, made just at the wrong moment).

Consultation: A management style in which consultation is inherent acts as the best foundation for an empowered way of operating. It means that the framework within which people take responsibility is not simply wished, perhaps seemingly unthinkingly, upon them, but is something they helped define – and of which they have taken ownership.

Feedback: Empowerment needs to maintain itself; actions taken must not sink into a rut and cease to be appropriate because time has passed and no one has considered the implications of change. Feedback may be only a manifestation of consultation, but some controls are also necessary. Certainly, the overall ethos must be one of dynamism, continuing to search for better and better ways to do things as a response to external changes in a dynamic, and competitive, world.

Development: It is axiomatic that, if people are to be empowered, they must be competent to execute the tasks required of them and do so well. Remember, too, that useful development is itself always a significant motivator.

An enlightened attitude to development is motivational. A well-trained team of people are better able to be empowered, for they have the confidence and the skills. An empowered and competent team are more likely to produce better productivity and performance. It is a virtuous circle.

8.7.11.4 Achieving the right balance

At the end of the day the answer is in your hands. Keep too tight a rein on people and they will no doubt perform, but they may lack the enthusiasm to excel. Management should have nothing less than excellence of performance as its aim. Market pressures mean that any other view risks making the organization vulnerable to events and competitive action.

On the other hand, too little control – an abdication of responsibility and control – also creates risk. In this case, staff will fly off at a tangent, losing sight of their objectives and, at worst, doing do more than what takes their fancy.

As with so much else, a balance is necessary. Empowerment is not a panacea, but an element of this philosophy can enhance the performance of most teams. Achievement and responsibility rank high as positive motivators, and empowerment embodies both. Motivation will always remain a matter of detail, with management seeking to obtain the most powerful cumulative impact from the sum total of their actions, while keeping the time and cost of so doing within sensible bounds.

> **Networking hint**
> Empowerment is one more arrow in the armoury of potential techniques available to you, but it is an important one. Incorporate it in what becomes the right mix of ideas and methods for you, your organization and people; make it clear to people how you operate and it can help make the whole team work effectively.

Finally here are ten keys to adopting a motivational style that summarize what makes for success.

1. Always think about the **people aspects** of everything.
2. Keep a list of possible **motivational actions**, large and small, in mind.

3. Monitor the **'motivational temperature'** regularly and often.
4. See the process as **continuous and cumulative**.
5. Ring the changes in terms of method to **maintain interest**.
6. **Do not be censorious** about what motivates others, either positively or negatively.
7. **Beware of panaceas** and easy options.
8. Make **sufficient time** for it.
9. Evaluate **what works best** within your group.
10. Remember that, in part at least, there should be **a 'fun' aspect** to work.

Make motivating, and the communication that transmits it, a habit. Take a creative approach to it and you may be surprised by what you can achieve with it. The motivation for you to motivate others is in the results.

8.7.12 Aiming for excellence

Finally in this chapter, remember that even the best performance can often be improved. Motivation is not simply about ensuring that what should happen happens. It is about striving for – and achieving – excellence. All sorts of things contribute, from the original calibre of the staff you recruit to the training you give them, but motivation may be the final spur that creates exceptional performance where there would otherwise only be satisfactory performance.

It is an effect worth seeking; and it is one multiplied by the number of staff involved. How much more can be achieved by ten, twenty or more people all trying just that bit harder, than can be achieved by one manager, however well intentioned, doing a bit more themselves? Motivation makes a real difference. If you can motivate people, your people skills are developing well.

Chapter 9
Results, referrals, rewards

This chapter puts all that has previously been explained together. You are now confident when you meet new people. You're able to make connections and develop powerful relationships. These will be beneficial, not only to you personally but also for your organization. You will be **reaping rewards** through **referrals** for your company, your staff and for yourself. This encompasses all aspects of business **relationships**, how to get a **response** to your efforts.

Because you have by now become more experienced in dealing with people and different situations within the workplace, you are far more confident when dealing with people who can affect your organization externally. When you are dealing with potentially valuable business connections, remember to switch to receive mode. If you had been transmitting, it's now time to stop. In this way favours can be returned. You'll feel enriched once you start creating new relationships that result in new business, and rewarding contacts.

Don't forget that networking/relationship building is a reciprocal exercise – what goes around, comes around. Referrals and recommendations will come to you frequently. Repeat business will be the norm.

> **Networking hint**
> Because you have worked at your networking, making connections and building relationships, your strategy is in place.

There are a few rules that should keep your relationship building on track.

9.1 Keep up the numbers

If you find your new contacts begin to decrease, you could be stuck in a routine. If you do what you've always done, you'll get what you've always got. People make decisions about where they spend their time based on perceived worth. If the networking events you've been attending have lost their value to you, take a step back and have a look at your options for change.

Maybe you could vary the organizations to which you belong, or the ways in which you seek to make new contacts. Make a 'hit list' of new people you want to meet, and try to contact between three and five each week. Build in a culture of asking your new contacts to introduce you to one or two new people and doing the same for them. This way your 'new blood' simply has to increase.

Speak to people who you have not contacted for over six months to a year. Follow up those who have left organizations and moved on to pastures new. Build relationships with their successors and track down your old contacts. They will be flattered you have traced them and be happy to re-establish a connection with you. Who knows? They may just have been waiting for you to contact them.

Ask friends and colleagues to invite you to visit their business networking events as a guest. It's sometimes a good idea to meet a completely new crowd. And what a great morale booster if you get there and someone recognizes you!

When breaking new ground, remember the **five rules** of relationship building.

9.1.1 Rule 1: Create empathy

This is the ability to put yourself in the other person's position and see things from their point of view. It may come naturally to you, or you may have to acquire this skill. Empathy is vital and it has to be visible. Your business contact must feel that you understand. When you hear him say, 'You're a good person to work with,' you can be assured you've got empathy!

To start the process, introduce yourself briefly and set about the task of finding some common ground as quickly as possible. It should be possible to establish two or three things in common with a new acquaintance within a minute or two.

Small talk should be used as a tool. The purpose is to uncover something that you have in common that will help establish rapport. Once you are no longer strangers, you have begun the process of establishing an individual connection. It is much easier to build a relationship once that stage has been reached.

9.1.2 Rule 2: Be courteous

Engage with someone by being sympathetic. It will surprise them and make them feel human. Small talk can often seem superficial and artificial. Get into real conversation with your business contact and watch for the warmth of their reaction. Look for visual and oral clues to assist in establishing the relationship. Make your voice warm and engaging and use positive body language. Watch for any signs of mirroring to help you.

9.1.3 Rule 3: Be enquiring

Use open questions to elicit information and encourage conversation. If you're having difficulty in eliciting information from someone, it can be very frustrating being faced with just yes-and-no answers. There is a technique called 'the string of pearls'. It means connecting one thought to another. You can try practising this technique during the small-talking process at business events. Openers, such as a book you've just read, or a film you've recently seen, can be sufficient to get you started.

9.1.4 Rule 4: Create interest

Keep an expression of interest in what you are saying. Be alert to the possibility of throwing in an unusual question or quick response. Sometimes humour is appropriate to maintain levels of attention. You can ask their opinion about something as a hook for making a comment. Don't forget to see the other person as someone of importance – put yourself in their shoes. If you can use this skill, you will 'humanize' your connection – after all, that's how you'd like them to be thinking of you.

9.1.5 Rule 5: Show respect

Don't assume that your business contact will have the same views and attitudes you have. The world is full of different people, all with varying ideas, values and opinions. These may not be similar to your own. Their culture may be very far removed from yours. But that does not mean it is less important. You may find some attitudes and customs unusual. Working practices may seem positively odd. In order to build a working relationship it is helpful to be able to get on with them.

Argument and confrontation are not the best basis for building rapport. If you can understand them, you may be able to respect them even if their attitudes are wildly different from your own. Respect your contact's individuality, and take an account of what 'type' of person you are dealing with. If you make a concerted effort to 'get on their wavelength' you will find this has a positive effect in subsequent dealings with them.

Be considerate of their views and values. This will indicate that you have an open mind and a caring manner. It will also convey respect for your contact's taste, views and attitudes. Your business contact will appreciate the attitude and respect this shows.

9.2 Make it real

9.2.1 Sincerity

Sincerity is one of the most influential factors in relationship building. No one is going to be interested in you if you're artificial. Be genuine. Know what you're talking about and mean what you say. If you're ignorant about the facts, whether it's to do with company policy, a particular area of expertise or specialist knowledge, you need to get out of the office and meet people who know the answers. Ask lots of questions and then listen.

9.2.2 Speak plainly, don't use jargon

Jargon excludes people, whether they're members of the general public or newcomers to your network or organization. The more you can avoid these expressions creeping into your conversation, the better response you will receive from your business contact.

9.2.3 Why should people respond to you?

Give people a reason why they should respond to you. This should be some kind of benefit.

If your business contact knows that you are able to organize deliveries at an earlier time in the morning, or collections later in the afternoon, they will be interested.

Don't make promises if you can't deliver. Consider whom you are dealing with: a donor, a user, a volunteer, a service provider? Depending on their role, their 'take' on the issue will be slightly different. Adjust your approach to reflect their interests.

If you are well informed and have a complementary interest, this should be a positive help. Say you are trying to win over an influential new contact. Find out what he is passionate about. If it is something you have some knowledge of, share it with him. Otherwise, do some research so that, when you meet next time, you can ask informed questions that he is most likely happy to answer.

People make decisions about others on the basis of both the rational and the emotional. Relationships are founded on certain values. Identify those values and make sure they are reciprocated. If they underpin everything you do, they will engage someone emotionally as well as intellectually.

9.2.4 Brand and company identity

Maintain brand identity where business relationships are concerned. You may not have the most expensive and impressive corporate house style or logo. But, if you are neat and cohesive in your approach, you will command interest and, hopefully, respect. Scrappy and unprofessional literature reflects poorly on your organization and the service it offers.

By maintaining high standards, personally and organizationally, your business contacts will recognize and respect your company's identity. Your reputation directly affects the likelihood of developing a successful corporate relationship. If you are going to build strong business relationships you are clearly motivated by a deep sense of value. You are proud of who you are and what you do. Delivering outstanding service is not just to win business.

You are developing relationships with other professionals who have the same sense of values. These are:

- honesty
- sincerity
- responsiveness
- confidence
- modesty
- trustworthiness
- appreciativeness.

9.3 Recommendations and referrals

Word of mouth is a powerful tool – using other people as ambassadors for you, or your company.

If you offer to recommend your contacts and help them with their business development, they will reciprocate. It is possible to build a business entirely on referrals with the right network in place. New business obtained on the recommendation of other people is highly profitable. The clients usually spend more and are loyal. But few companies actively seek referrals.

9.3.1 Pareto's Law

The nineteenth-century Italian mathematician, Pareto, gave his name to Pareto's Law. It is often known as the 80/20 rule. It suggests that 80 per cent of your business comes from about 20 per cent of your clients (as mentioned earlier in Chapter 5).

A recent survey showed that almost 80 per cent of a company's customers would be willing to act as referrers for them, but only 20 per cent of them were asked to do so. A company that asks for referrals from 80 per cent of its existing clients would increase its annual turnover by 20 per cent.

Take time to organize and categorize your business contacts and use this arrangement to help you work out how you deal with them, how often you contact them.

9.3.2 Why recommendations and referrals work

Recommendations and referrals work because they come from a trusted source that has already benefited from the commercial or professional relationship yet has no vested interest in the business.

Such endorsement is independent and unsolicited. In business terms it has a rapid conversion rate – acceptance being dramatically accelerated because the service or product has already been tried and tested by a reputable third party.

When building business relationships you are setting up the perfect mechanism for profiting by referrals. The key principles are 'giving to receive'. Satisfied customers are both loyal and profitable. If you are always on the lookout for opportunities to recommend products or services that you trust and admire, you can expect this to work in your favour.

Make it a rule, therefore, when beginning a business relationship to ask how you can help them. What organizations or people do you know who could benefit from their services? If you build a 'rapport web', it can be an ever-increasing source of new business. Consider your business network's own contacts database, and so on. It's like ripples on a pond – self-perpetuating if you are persistent.

Don't be afraid to be direct. Make sure anyone who can offer a referral knows as much as possible about you or your company's services. If they do not understand the value of your product or services, or appreciate the key benefits, they will not realize what makes you outstanding and memorable. This is the message that they should be communicating.

Referrals are the rewards that come once you have created your network, made your connections and built your relationships. They are the ripe fruit that you have worked hard to grow. Like a harvest, they only appear if the conditions are favourable (i.e. deserved).

The time to ask for a referral is when:

- you've introduced someone to one of your contacts who awards them a project;
- you have successfully completed at least two transactions for a client;

- you solve a problem for someone who wants to reciprocate in some way;
- someone thanks you for providing a good service;
- you've helped someone through a particular difficulty;

> **Networking hint**
> Asking for referrals is one of the most powerful and low-cost ways of building or developing your career or business. It is a simple approach which feeds on its own success. But it has to be built on secure foundations.

If you are leaving referrals to chance, and thinking of them only as an occasional welcome surprise, you are ignoring an essential part of your business-development plan. Someone who knows that you and your company are genuine and is prepared to pass on that information to an interested third party is worth their weight in gold.

When you were taking those first tentative steps towards developing your interpersonal skills, how did you feel? Nervous? Ill at ease? What about the occasion when your contact said, 'Oh, you must know my friend X?' Hearing a name that is familiar encourages you to respond positively in a given situation.

A referral works when the relationship is all about trust and respect. Business success is built on the quality of relationships we have with others. Relationships come in various and subtle shapes and sizes. Never assume you know what role a person has in an organization.

Inquire, investigate and direct your attentions appropriately. Collaborative relationships are the ones that you're attempting to build, so you need to be knowledgeable as to how this is going to happen.

9.4 Every relationship is different

Without business contacts, preferably positive ones, you will make little progress. All your contacts are unique, some are demanding, others quite difficult, and some require understanding. Understanding them doesn't just make your relationship building easier: it is inherent to the whole process. Where there is no understanding, how can a business relationship develop?

You must make sure you know sufficient about your business contacts to enable you to do a good job of rapport building when you are with them.

If not, you simply will not be taken seriously and you will lose credibility. Make sure you have background knowledge, and that it is noted somewhere. Add other information as you glean it – this could relate to your contact's current circumstances (a forthcoming marriage, say, or an anticipated promotion). Accurate background information coupled with an empathetic approach will ensure that you have a winning formula.

Remember that your success in building brilliant business connections relates directly to your understanding of the other person – make it your business to know as much about them as possible. That makes it much easier to build great rapport.

9.5 Advanced relationship-building skills

By whatever means you attempt to put across your case, you should try to set yourself apart from your other competitors or contenders in the relationship-building process. It is only by being memorable and outstanding that you will be preferred.

You should bear in mind that other people are attempting to reach a similar position to yours. If you have researched and found that a particular person is an ideal business contact, it is safe to assume that others will have done the same. Assume that they are professional, as you are.

> **Networking hint**
> Whatever else needs to be done, however you try to exert influence over the relationship-building process, you need to be persuasive in your dealings with your potential clients, influencers and recommenders.

So how do you get to be 'first among equals'? The answer: become a persuader.

Building successful business relationships is a complex process. Contacts want to decide 'their' way whether or not to do business with you. They want to think about the proposition you are making to them, to assess it and make what they would regard as a considered decision. Their thinking is made up from many differing viewpoints. If you imagine there are two identical cases on a set of scales, how do you end up on the winning side?

To be an effective persuader you should be able to communicate to your contact that:

- he is important and will be treated as such;
- his opinions and position are respected;
- he will be dealt with as a unique individual;
- the benefits that are made available to him by dealing with you;
- what the facts are;
- any snags there are (and there usually are some);
- what compromises will be required;
- how the relationship can work.

9.5.1 What defines 'persuasive'

As an effective persuader, you need to ensure that your approach is seen by your chosen business contact as:

- understandable
- attractive
- convincing.

None of these on its own is enough to secure a successful business relationship. They need to be strong enough jointly to set you ahead of any other parties trying to negotiate arrangements with your contact. To be convincing and effective, your approach to rapport building must be individually tailored to the other party involved in the process.

Each business contact wants an approach that they see respects their point of view, matches their personality and interests and so generates more immediate response and interest.

Bearing this in mind will get you off on the right track. It should quickly show you the reason why the identification and preparation stages are so important. This is complex and there are many things to consider, including your own personal positioning.

If handled smoothly, the relationship-building process will appear well thought out and relevant. Your contact will know that he is dealing with a professional who takes time and care over each contact he makes and treats each one with respect.

9.5.2 Make what you say attractive

Communicate clearly, and make your words attractive so that your business contact wants to listen and is as keen to develop the relationship as you are. How do you do this?

9.5.3 Talk about the benefits

People do not encourage relationships in business just for friendship. It has to go further than that. They will want a clearly defined purpose. If you explain the reasons why you want to connect with them and that there are advantages to you both, they will understand 'what's in it for them'.

The advantages, say, of a small specialist design group collaborating with a large architectural practice means that the small firm can be included in bigger projects than they would normally get involved in. The large company will harness external specialist expertise in an area they do not have covered in-house.

Working together to bid for significant projects will result in a win–win solution. Talking of benefits in this way as you describe your ideas for increased working opportunities will support the relationship-building process. It is important to get this right. All contacts are different and in some cases you will be persuading more than one person of the advantages of such an alliance. For example, you could be required to influence a board of directors, or a group of partners in a professional-services firm. Communicate with everyone, respect their opinion, value their contribution, gain from the experience.

9.5.4 Make it credible

In defining persuasion, it is important to make what you say to your business contact credible. Most people who are experienced in business dealings have a healthy degree of scepticism. They can be forgiven for thinking that you have a vested interest and will be looking for an element of 'proof'.

The main form of evidence has to be the persuasiveness of the case put forward, harnessed to the tangible business benefits, followed by proof positive that it can be done. Suppose you were able to report that, 'Two years ago we collaborated with a company, X, who were looking to expand in Europe. Because of our strong associations in France, Germany and Spain, we were able to open up new markets for X in these countries.' This would show your new potential partner that you have a proven track record that backs up the proposal.

Such evidence would help to convince even the most cynical business contact. There is factual, physical proof here. Your contact can go and check the record and be reassured that what you say is true and not a fabricated claim.

If you find that your claim needs further substantiation, it may be helpful if there are outside elements that can be utilized. Perhaps your professional association has written a report about the achievements made by you and your previous partner firm. There may be other independent parties who are aware of your successful alliance. Maybe you made a presentation to another organization, or wrote a paper for a professional journal.

There are a number of ways external proof can be harnessed. These independent authorities are powerful persuasive elements in building credibility with your proposed partner. You can probably think of other examples, applied to your own company or area of expertise. Whatever your profession, industry or sector, assemble all possible independent proof factors for appropriate use when required.

9.5.5 Add value

Adding value will depend very much on your individual expertise or company policy. However, where possible, try to offer:

- more than usual;
- more than the competition;
- more than expected.

This could have irresistible appeal to your business contact. You could, for instance, suggest that you have a trial period of three months of the service; an incentive; an aspect of the collaboration that you can offer *pro bono* – particularly in relation to a charity or not-for-profit organization.

Any such device as this can act in a number of different ways to:

- help you get a better hearing;
- help improve the weight of the case you can present;
- persuade people to act now rather than later.

However you choose to persuade people of the benefits of your business relationship, this can only be part of your organized rapport-building strategy. Some suggestions will work better than others. You may be able to control parts, but you will not be able to control all of the process.

9.5.6 The fun factor

Use humour to improve creativity and lower stress. In your dealings with other people, the ability to use humour can work wonders. It aids communication, establishes empathy, diffuses awkward situations – and even builds the bottom line. Studies show that humour can increase productivity because it:

- increases the immune system's activity;
- decreases stress hormones, which constrict blood vessels;
- increases the antibody immunoglobulin A.

If you use or experience positive humour it involves the whole brain, not just one side. The result is better coordination between both sides. This means you are more relaxed, your blood pressure and heart rate are lowered and you are able to think more clearly. If those are the benefits to you, imagine how persuasive you will sound to your business contact.

Keeping a good sense of humour shows an ability to be relaxed. You can often get an impression of whether humour is appreciated by looking around you. If you are in someone's office, for instance, are there any amusing signs, cartoons, slogans or pictures? Do other people seem relaxed and able to joke with each other? When people have to decide whether or not to work or collaborate with you, they will be influenced by how they feel about you. By including humour in your dealings with other people, you are encouraging them to like you.

Laughter reduces stress because it is relaxing and calming. It has been shown in hospitals that patients who have had 'humour therapy' recover more quickly from illnesses or surgery than those who do not laugh. If you are trying to build rapport within your organization, the first time a new employee laughs at an 'inside joke' shows that she's part of a team with her co-workers. You know you have an 'inside joke' when everybody from a group laughs but no one outside the group does.

How many times have you noticed that when you are stressed you fumble, drop things or make mistakes? It is a myth that laughter is trivial. On the contrary, it is very powerful. Even just smiling can be healing and reassuring.

9.5.7 The personal touch

Personal connections can increase the bottom line. Try to develop your own skills with regard to relationship building throughout your company's network. The better-thought-out, innovative and sincere relationships are likely to be more profitable. You should put equal emphasis on your internal relationships with colleagues, staff and superiors as you do with the external ones – existing and new customers, suppliers and referrers.

> **Networking hint**
> Harnessing the power of personal connections builds strong and lasting relationships. You are looking for ways to match people, organizations and opportunities.

What will help you do this? Here are a few T words to consider. Establish **trust** – respect your business contacts and your staff. They will respond in a positive way.

Remember to say **thank you** – it costs you nothing and gains you much. If someone says '**Thank goodness** we met!' – **treasure** that moment. You may get **testimonials** from satisfied clients or recommenders. When other people blow your **trumpet** for you, the sound is much louder. You will find your **thinking** processes have changed. You will be far more people-oriented – whether externally or internally. With **tenacity** these relationships will flourish.

9.5.8 Hold on to your assets

Before moving to further advanced methods of dealing with external business contacts, let's look inside the company for a moment. Many organizations know that staff hold the key to success but few realize how to capitalize on their greatest asset. So often, companies focus on meeting customer needs or increasing the bottom line and forget that it's employees who can hold the key to success.

Good practice for people management is a variable process. Some of the main factors involved here are:

- enlightened leadership culture;
- staff involvement at every level in the organization;
- staff development to make employees feel valued and challenged;
- flexible work patterns to meet the needs of the business and staff.

If you knew it was going to cost you three times as much to replace a member of staff than to retain one, what would you do? Not exactly a difficult question. But how many people really pay attention to this? When a company loses a key member of staff, it is often the last to realize that it would have been much cheaper to tackle the issues that made that person quit than go through the expensive steps needed to replace her.

The real impact on businesses that lose staff regularly and are constantly recruiting new ones is that it probably takes a year's salary to arrive at 'break-even' point. Consider the case of a senior manager who leaves after, say, two years' service. Replacing him, the company will pay a recruitment agency, allocate HR administrators and involve directors' time in interviews. The two years of training investment will have disappeared. There will be loss of productivity in the run-up to his departure. His staff will be demotivated and, until the replacement is up to speed, may lose momentum and direction.

Above all, and this is where the real cost lies, all his business relationships will disappear with him. When you consider the reasons why he resigned – if indeed you ever know them – would it not have been cheaper and easier to pay attention to those issues and retain the manager? Was it a salary factor? Was it his boss? Perhaps he felt undervalued.

Whatever the solution – and it may not have been simple – it would have been cheaper to address those issues than to replace that key member of staff. If you value your staff as assets, not just overheads, and retain them and their business relationships, you will save your company significant amounts of money.

9.5.9 Realize their potential

Let's leave the financial issues of staffing for a moment. The subject is *valuing* staff, *maximizing* their potential and regarding them as *tangible assets*.

> **Networking hint**
> People can, and do, make an enormous difference to any organization. What motivates and drives them to release their potential is the way they are treated by their company.

The best way of building good relationships with staff is by finding creative ways of motivating them, sharing knowledge with them for the benefit of the whole organization. The skills of motivating staff and delegating tasks to them were dealt with in detail in Chapter 8 (see Section 8.7) and to a smaller extent elsewhere. Bringing emphasis on creativity through the redesign of work areas or working conditions can make a huge impact on staff morale. Happy staff will remain with the company. If HR and recruitment costs drop, the increase in the company's bottom line is significant.

9.5.10 Giving recognition for good service

Developing a 'praise culture' is something some enlightened firms practise. This can be applied in a number of ways: public praise for good work, support for entry to industry award competitions and more social opportunities. Positive moves, including new training methods, cutting-edge technology and introducing a profit-sharing scheme for individual staff on the basis of performance, could be other winning formulae for improved staff retention.

Organizations that practise good people management find it brings many benefits, including financial success and retention of key staff and their business connections.

9.5.11 External relationships

One of the most effective ways of harnessing the power of personal connections in an external way is to use it to develop your business. You can extend the relationship-building process for the specific purpose of increasing turnover by simply *enquiring* and *listening*. What does this mean? In essence, a customer-satisfaction survey.

> **Networking hint**
> It costs a quarter of the amount to retain an existing client than it does to win a new one. If you can keep your clients happy, you will make business development easier and more successful.

Why is it important to find out customer feedback? Because it is an essential piece of management information. For professional-services firms, customer feedback can be a remarkably inexpensive source of market research. You can find out:

- who your customers are;
- if they are not your customers already, then when they are likely to be;
- why they are your customers and not someone else's;
- what your customers want;
- how your customers feel;
- what your customers think;
- how you can make your customers feel valued;
- what sort of initiatives would your customers appreciate;
- what you can do to keep your customers loyal;
- how you can give yourself a competitive edge over others.

There is a huge amount of information here and it should be available for use by the management structure to increase turnover and improve the bottom line.

Many successful companies use this method to enhance corporate connections and reinforce existing business relationships. You may not need to answer all these questions but you should take the trouble to find out as much as you can about the psychology of your customers. If you don't know what's happening, you won't know how to deal with situations when they occur.

If you take the time to talk to people who already do business with you, who have paid money for your services, you will be asking their advice, which is flattering. In turn, they will feel that their opinions matter. They are less likely to desert you in times of difficulty if you have made them feel important.

Monitoring customer satisfaction is a pointless activity unless management have a sense of ownership for the process and are prepared to act on the results.

One company succeeded in increasing their annual turnover by up to one-fifth, by managing a tailored approach. The objective of the survey has to be defined before the programme begins, together with a budget and a timetable.

Before embarking on such a survey, it is useful to review existing information or research data concerning customers and customer satisfaction. You could ask the following questions.

- What do I know about our existing customers?
- What do I know about their expectations?
- How well are we meeting those expectations?
- What will happen in the future to customer requirements?
- How do we compare to our competitors?
- How is the market likely to change in the next three years?

Case Study: Brand values

Following a rebranding exercise three years ago, Company X decided to carry out some empirical research to assess how the new brand values and perceptions of the business generally had changed/improved. This was conducted by means of a survey of their clients, potential clients and market influencers.

The research was designed to give a clear indication of the competitive market position that Company X currently occupied and point to the effectiveness of the company's different services and market sectors. It also set out to show how they fitted with the image, identity and reputation that it projects overall.

In addition, it highlighted the relative strengths and weaknesses as observed by a representative cross section, whose opinions were specific and insightful. It offered opportunities for the future as well as areas where increased marketing activities would produce a measurably greater return on investment.

The most important result of the survey was the influence it brought to the firm's strategy for building further value into their brand for the future. It helped clarify issues surrounding the way to create best market differentiation so that it could compete even more successfully with its known competitors.

Of the three categories of influences, the research asked questions in the following areas.

Existing clients:

- reasons and motivations for choosing Firm X in the first place;
- satisfaction levels in the relationship;
- satisfaction with final results;
- how Company X's character is perceived: innovative, conservative, in tune with needs, prestigious, dependable, friendly, formal, etc;
- relative strengths compared with others in the marketplace;
- relative weaknesses compared with others in the marketplace;
- extent to which they would refer business and for what specific product/service/application;
- awareness and effect of the new branding;
- relative strengths of different market-facing divisions.

Prospective clients:

- familiarity with, and knowledge about, Company X together with perceptions of their relative market positioning;

Case Study: Brand values (cont.)

- values associated with the brand (plus any historical perceptions);
- factors that most influenced choice – referral, reputation, track record, the quality of individuals, relevance of previous projects;
- who's who in the marketplace and their relative strengths (weaknesses?);
- expectations of service performance;
- expectations of individuals;
- extent to which Company X could/would be referred to others.

Market influencers/referrers

- extent and nature of reputation;
- effect of the rebranding;
- quality of the people;
- quality of service/results;
- specific areas where Company X would be referred;
- specific areas where it probably would not;
- relevance/value of different product areas;
- suggestions for product/service development.

Method

The directors of Company X decided the specific areas and outcomes required from the research survey questions. They developed a questionnaire for each of the three categories above and included a mix of closed and open-ended questions to capture options and perceptions.

From their database they selected and approached a similar number of companies from each of the three constituencies. There was a need to modify the questions after one or two phone calls had not produced the desired results.

After completing the survey, they analysed the results. The report showed the analysis, drew conclusions and made recommendations from what the contacts and clients had said.

Conclusions

The overall results were positive and showed a universally high regard for Company X's professionalism and confidence in its ability to deliver consistently well-above-average results. The principles were considered to be approachable and well able to handle any difficulties in the relationships between Company X and their clients.

Case Study: Brand values (cont.)

It indicated that Company X was successful in its relationship-building events, and it has a reputation for quality events even from people who had never attended one!

It was recognized that Company X had been increasing its marketing and PR presence to maintain a modern image while at the same time raising its profile.

In common with many professional-services firms, Company X's directors were not only the principal practitioners of the work in terms of fee earners but were also required to manage the business. There was a consistently high regard for Company X's professional capability but there was room for improvement in terms of their business relationship building.

It was evident from the research that contact databases would benefit from regular updating. There were inconsistencies in terms of a business-development strategy and were Company X to effect a more integrated approach this would be beneficial.

Successful outcomes

Six months later, Company X were delighted to announce that, following the creation of the post of business development director, they had won a significant number of new projects. The results were tangible, applying an integrated and consistent approach to business-relationship building, Company X's annual turnover had increased by 19 per cent.

If you take that one stage further, to increase a company's annual turnover by almost 20 per cent by means of a customer-satisfaction survey, a small-to-medium-sized company with an annual turnover of £2 million would add on an extra £400,000 in a year.

The results could be even greater, so how can anyone afford to ignore it?

9.6 Further ways to develop professional relationships

9.6.1 Competitor analysis

As a general rule, keeping an eye on the opposition is a good way to make sure your organisation stays ahead of the game. Why? Because, at the very least, it's an information gathering process about the current state of the market. At best, it is a means of developing a niche area or minimizing the risks that can affect your own organisation. Some people argue that it is better to be 'first' than 'best'.

The well-known scenario about the tiger hunters in the jungle puts this important issue in context. There were two hunters in Africa, as they were emerging from the trees suddenly coming towards them they spotted a tiger. One of the hunters immediately bent down and started to put on his trainers. 'What on earth are you doing?' asked his companion. 'You'll never out-run that tiger.' 'I don't need to run faster than the tiger,' the first hunter replied. 'All I have to do is to run faster than you.'

It makes sense, doesn't it? There's no need to put the competition out of business. Just having an edge over one or two will be enough to enable you to thrive. But first of all you need to know who these people are.

Competitors are companies offering similar products or services to you; or companies offering the same products or services. They could be businesses who might offer the same or similar products or services in the future. Alternatively they could be organisations who will remove the need for such products or services as your organisation provides. Whoever they are, their objectives are the same as yours – to grow, make money and succeed.

'Never underestimate the enemy.' Wise words, but often ignored. To gain advantage over the competition means knowing how he thinks, how he might act, what his strengths are, where his weaknesses lie. Also knowing when and how he is vulnerable, where he can be attacked and knowing when the risk of attack is too great.

9.6.1.1 How to do it

There are five key stages to help you achieve this.

- *Work out*

Why do you need this information? What do you want to find out? How are you going to do it? Who will analyse the data? How will you use the information once you've collected it? What results do you want to achieve having got it?

- *Find out*

Who are your nearest three direct competitors? Who would you regard as indirect competitors? Which of these organisations is growing, static or declining? What can you learn from their operation and advertising? How would you describe their strengths and weaknesses? What differentiates your business (products or services) from theirs?

Don't forget that in normal economic times markets are constantly changing – legally (i.e. regulations and statutes), politically and in terms of technology. In order to thrive when there are other economic factors at work, any organisation needs to be able to adapt quickly to suit current trends and reap any possible benefits.

- *Gather information*

Carry out your research – look, listen and learn. Visit your competitors' locations, to observe how they do business, set out their products, offer their services. How do their staff treat customers?

One way of doing this is called 'mystery shopping' – market research companies use this method to measure quality of service or gather specific information about products and services. But there's absolutely no reason why you can't do this yourself (unless of course your face is well known to the opposition). If you delegate the job to a third party, this person must pose as a normal customer purchasing a product, asking questions, registering a complaint or behaving in whatever way you direct. They then report back to you with detailed feedback on what transpired. (Think *Fawlty Towers* – the episode of the hotel inspector.)

You could ask your own customers their opinion about such organisations. Keep an eye on competitors' marketing and advertising. Who are their target audience and what percentage of market share do they hold? Visit trade shows and exhibitions attended by them. If possible, go along to presentations or speeches given by members of their staff. Observe what appears in print about them – in professional journals, the business press, local newspapers and trade association publications.

- *Process the data*

Study and analyse the findings. What should become apparent are trends and patterns. These should be related to your organisation's development, profitability and market positioning.

- *Report findings*

Set up a system for evaluating the results. Feedback is essential for everyone involved. This means the information gatherers, the processors and the decision makers. Ask the questions: Was the information useful? Was it understood? How was it interpreted? What was the result of its use? Was it worth it?

Remember, the value of knowledge is difficult to calculate. You can't be sure how or when you are going to use it. But because you are trying to develop good strategic relationships and expand your business or market share, you can be sure that ignorance is far more costly. It can result in missed opportunities, or loss of customers. At the greatest extreme, the business itself could fail as a result of inactivity in certain crucial areas. Don't lose out, keep your antennae tuned. Competitive intelligence is a secret weapon you cannot afford to be without.

9.6.2 Partnering – strategic alliances

Sometimes, someone you think is your enemy can turn out to be your friend. It is always worth investing a little time in examining what is going on outside your organisation. It's easy to overlook the competition as a resource, but all you have to do is look at things in a different way. If you shift your focus to view your competitors as an addition to your supply chain rather than a rival to it, you will discover opportunities that would otherwise remain unknown to you.

Thinking of competitors as allies rather than opponents is not new. Strategic alliances are often the way forward for small to medium sized organisations. Cooperation with competitors, customers, suppliers and companies producing complementary products can expand markets and lead to the formation of new business relationships. In some cases it can create new forms of enterprise.

Networking hint
The idea of teaming up with competitors to develop new ideas and to make your organisation better at what is does, delivers a challenge to many people. But there are a number of ways of doing this.

Strategic alliances can be formal and encompass only a specific project. At other times they are informative and active with only certain types of project, as shown by the list below.

- *Development or extension of products or services*

If your business is customer-focused, you will actively seek out the best ideas and ways of serving your customers' needs. But combined strengths can produce amazing results. By collaborating with a competitor you might be able to win new contracts neither of you could do alone. One plus one can often equal much more than two.

- *Apportioning referrals*

Consider having at least three people or companies to whom you would refer business without hesitation. Ideally there should be a mutual understanding that the favour will be returned. Whether you have arranged a referral fee, a reciprocal referral, or you are the one that wins the project, everyone is a winner.

- *Get in the know*

It is essential to find out which are the best organisations producing complementary or related services in your own market. Knowledge is power and if your customers perceive you as the place to go for information, your business reputation will grow. Your customers will value your intelligence and connections in the market.

- *Best practice*

You can learn something from everyone and every situation. No one can possibly have all the answers, which is why sharing best practice is so important. It does not mean sharing trade secrets or colluding on fees. What it means is coming together for improvement. True professionals subscribe to the principle of abundance and see the power of helping each other to get better. 'A rising tide lifts all boats.'

- *Risk awareness*

It is important to bear in mind the possible pitfalls when contemplating strategic alliances with a competitor, or anyone else. One possible issue is the lack of common goals amongst the parties. If the collaboration does not work, perhaps the synergies were not real or the communication system was flawed. It is wise to do some research before committing yourself to such an alliance – a corporate version of the pre-nuptial agreement.

The obvious benefits of strategic alliances mostly outweigh the risks. It is important to pay attention to whether you really can work together. Complementary areas of expertise are one thing, but do the personality types fit together? The question to ask is, 'Can they really add value to the project?'

Creating successful strategic alliances is a valuable skill to acquire. You need to have complete awareness of your own strengths and weaknesses, as well as those of your company. Look for complementary strengths in your competitor cum ally. Ideally, it will be someone who actually makes you look better at what you do.

Chapter 10
Your checklists for success

You may be wondering where all this networking, connecting and relationship building is taking you. The answer to that is, you should know. It's your plan after all. Perhaps you should regard the process as a 'design-it-yourself' course in personal and career development.

If that makes it clearer, the major questions you need to consider are divided into these sections.

Current job/where you are now:

- How challenging is your job?
- Does your job/career path have a future?
- Does your work allow you to express yourself and your values?
- What skills would it be wise to upgrade in order to ensure career progression?
- Is anything standing in the way of your work success today?
- If so, what is it, and how will you deal positively with it?

Looking ahead/career management:

- Where do you want to be in your career in five years' time?
- Brainstorm and mind-map – decide what you want to be doing.
- Consider, if you had a choice, how you would get there.
- Plan your route.
- When will you start to implement the plan?

Coping with changes

- Can you cope with continual change processes?
- Do you find change processes threatening or challenging?
- Are the change processes happening around you positive ones?
- How will you manage those changes for your continued success?
- Are they better for you, or better for the company (ideally they should be both)?

When changes are imposed on you, you have no choice but to accept them. What you can do, however, is choose how you respond to those changes. If you can develop an adapt-and-thrive approach to changes, you will find that you can manage them positively.

> **Networking hint**
> If you think your networking strategy is having a positive influence, carry on. If you don't think it's working, don't waste your energy and time: redesign and rework it.

If you know that change is about to happen, it helps to be proactive. Get involved, prepare for the challenges that these changes will bring. By welcoming change you will find that you are able to take advantage of new and better opportunities. This could mean looking at ways of improving your skills, position in the company, relationships in the workplace, areas of responsibility and accountability.

If you can keep up with the changes that surround you and use them as a positive and motivational driver, they will help you achieve your goals.

10.1 Future planning

- Decide on your route to success.
- Plan your route – what do you need to do, whom do you need to have with you as part of your team?
- Acknowledge every step of your personal achievement.
- Continually motivate yourself.
- Set goals – sometimes goals can be achieved more quickly by involving others.
- Ask who around you needs to be involved.
- Take ownership and full responsibility for your targets.

Personal development, helped along by means of your business connections (mentors/influencers), should be a smooth process. Embracing change takes courage, because it involves leaving a situation you are comfortable with. You have to make a conscious decision to step into the unknown and welcome new challenges. But you know, from having developed a vibrant network of contacts, that there is always someone you can turn to for advice, support and encouragement.

10.2 Mentoring

Mentoring is something that could be regarded as part of the relationship-building process. A mentor is someone who exercises a low-key and informal developmental role (I explain more of the basics of mentoring in 10.2.2 below). During your networking you could have found someone who would make the perfect mentor for your personal and professional development. Personal development is not an attractive luxury: it is an ongoing necessity. To succeed in any organization you need to change and adapt along with the circumstances.

Personal development does not just happen. No individual employee can assume that development will be automatically provided by their employer, certainly not all that is desirable. Everyone benefits from taking an active approach to development.

Networking hint
Mentoring is the technique of creating a relationship, usually separate from normal reporting lines, where one person helps another and does so on a continuing basis, with much of what happens conducted on an informal basis.

10.2.1 Finding your mentor

One of the motivations for being a manager, or at least for some managers, is the satisfaction of helping people develop and of seeing them do well. Some people are lucky in that, early on in their careers, people they worked for took on the role of mentor to them. Today, they owe much to these people. It is something well worth aspiring to – whether to mentor others (if you are in the position of giving something back) or seeking a mentor from among those with whom you come into contact. At best you learn a great deal and very much more quickly than would otherwise be the case. On the other hand, luck does play a part here. There may not be suitable candidates around; not everyone has the courage or opportunity to seek out such assistance, particularly early on in their careers.

The ideal mentor is sufficiently senior to have knowledge, experience and clout in your chosen profession. They need to believe in the process, for it to appeal to them, and they need to have time to put into the process. Such time need not be great. The key thing is to have the willingness to spend some time regularly helping someone else. If you are keen on developing a relationship with a mentor, there is no reason why you cannot have regular contact with a number of people where, in each case, the relationship is of this nature. This can take various forms depending on the area of expertise of the individual. For example, if you wish to benefit from a number of different areas of expertise, you may have to seek out a number of different individuals. Such an arrangement is not uncommon.

10.2.2 Mentoring explained

Perhaps here is a good place for a little explanation as to exactly how mentoring works. It can be so useful. As I said earlier, a mentor is a person who will exercise a low-key and informal developmental role. More than one person can be involved in the mentoring of a single individual. While what they do is akin to some of the things a line manager should do, as has been said, more typically, in terms of how the word is used, a mentor is specifically *not* your line manager. It might be someone more senior, someone on the same level or from elsewhere in the organization. An effective mentor can be a powerful force in your professional development.

What makes a good mentor? The person must have authority (this might mean they are senior, or just that they are capable and confident). They should have suitable knowledge and experience, counselling skills, appropriate clout and a willingness to spend some time with you (their doing this with others may be a positive sign). Finding that time may be a challenge. One way to minimize that problem is to organize mentoring on a swap basis: someone agrees to help you and you line up your own manager to help them, or one of their people.

Then a series of informal meetings can result, together creating a thread of activity through the operational activity. The topic of these encounters may be general – a series of things all geared, let's say, to improving the calibre of your external business relationships. Ultimately, what makes this process useful is the commitment and quality of the mentor. Where such relationships can be set up, and where they work well, they add a powerful dimension to the ongoing cycle of development.

This is really much more than just networking. But it is an advanced form of the process, designed to help with personal and career development. The nature and depth of the interaction and the time and regularity of it is much more extensive. This is not primarily a career-assistance process in the sense of someone who will give you a leg-up in the organization through recommendation or lobbying, though this can of course occur. It is more important in helping develop the range and depth of your competences (specifically in relation to improving your people skills), with this in turn acting to boost your career. A senior or well-connected mentor may also act as an early-warning system when trouble is brewing.

As a final note, the perfect mentor is someone who makes an art form of this kind of role. Mentoring is a development technique that can assuredly make a difference, and a very positive one at that. Job performance and career progress can both benefit directly.

10.3 Build on success

Ongoing success involves cycles of activity:

- understanding the key things that can create success;
- being conscious of how you do things as you do them;
- planning and acting in accordance with that;
- monitoring the results arising from what you do;
- fine-tuning, and building in the experience of how things work to improve what you do next.

It is worth noting that an effective networking strategy is cyclical. Your initial networking helps you to connect with the right people. Once you have done that you are in a position to develop profitable business relationships. Such a good start is always desirable in its own right. But, as you develop new relationships, this will in turn bring you into contact with more new people and so the process continually evolves. Pay particular attention to your early successes in the process. This will influence how you take things forward for the future.

From the beginning, promise yourself that you will be a great networker and build good rapport with a variety of people.

- First, organize yourself and then you can work on other people.
- Take your time (you can't achieve miracles in a day).

- Make the effort (no quick fixes or magic formulae).
- Keep thinking (the obvious or immediate answer may not be the best).
- It's not a solo effort (seek and take advice from wherever you can).
- It will not always go right (admit your mistakes, publicly if necessary).
- Learn from your experiences.
- Keep good, regular, open communication.

> **Networking hint**
> The opportunity of getting off to a good start may occur only once, but the effect it has is long lived.

Being an effective networker is not so much what you do, but how you do it. You need to be able to engage people in the process. In the end, success comes down to a considered approach. Charge in, desperate to make an impression, go at everything at once and disaster may follow.

Taking on board the suggestions throughout the book will not make the networking effortless. However, by reaching this final chapter you will have formulated your own impression on how you are going to set about the process. So you need to make as good a first impression as possible when starting off. This can be reinforced the more people get to know you and the more experienced you become. There is much that can aid the learning process, but only one person can guarantee that you will succeed in your relationship building – you. Using the advice in this book, you could hardly do better.

10.4 The story so far . . .

This book was designed to take you from stage to stage in the networking process. In the beginning the purpose of networking was explained. The continuation of the process included making connections in order that profitable business relationships could be developed. The chapters dealt with the various aspects in a logical sequence.

By way of summary, here is a review of what areas have been covered.

10.4.1 Who needs business connections anyway?

Everyone needs them, particularly people who are ambitious and want to succeed in their career. Remember, task awareness is fine, but being proactive about people will take you further faster. In short, people mean business.

There are two main reasons why you should harness the power of personal connections: (a) your company will be more successful and will stand out ahead of its competitors; (b) you will progress further and faster along your career path than someone who doesn't actively network. Personal recommendations speak volumes and are more impressive than the best CV.

10.4.2 Business success is 20 per cent strategy, 80 per cent people

Developing successful relationships at work means looking in two directions: internally (within the organization or profession) and externally (among clients, work providers, suppliers).

In both cases, it helps to be confident and to cultivate extrovert characteristics.

10.4.3 How to distinguish between networking, connecting and relationship building

As a means of generating business or career progression, remember that a strategic networking plan is essential. From this, you should have the ability to establish unique personal and professional connections. These connections should be nurtured so that valuable business relationships are developed.

Individuals should each make the most of their own personal strengths. Make sure your plan is appropriate to your needs and that it connects with your company's marketing and business development strategy.

> **Networking hint**
> There are certain key attitudes and actions you can use to maximize success. It is not important what system you adopt as long as it works for you.

Here is a short list of some of the qualities needed to build good rapport with business connections:

- believe in yourself;
- believe in your company;
- meet lots of people;
- listen to your new business contacts;
- maintain a sense of humour;
- revisit existing contacts regularly and often;
- offer and accept referrals from everyone you meet;
- follow up on every connection made;
- ensure that the relationship is reciprocal.

10.4.4 Remember that R is for Relationships – and more

Relationships – why build them? Two reasons: they're practical and profitable. The most successful people are the best connected.

Recognition – creating impact when you meet people. Never underestimate the power of first impressions.

Recall – if someone can recall you easily to mind, you made (hopefully) a favourable impression when you first were introduced.

Reaction – one of the things you are hoping for is a positive reaction when you encounter them again.

Respect – aim to gain their trust. The ability to cooperate with and assist others is vital. You will then earn respect. Don't forget to show it to others in return.

Responsibility – you should take responsibility for your business relationships. That will keep you in control of your personal network. It's worth a lot to you – don't let others mess it up.

10.4.5 Why contacts are so useful

Everyone has their own group of personal contacts – their unique network. How many people are in yours? Do you value it? How do you use it?

Some ways in which it can be used are as a **research aid**, for information gathering; as a **link to new clients** or markets; and to **advance your career** by seeking influential people.

Here are some important stages in the business relationships process. Strategic networking requires a number of action steps.

10.4.5.1 Review your existing contacts and connections

- How up to date are these?
- Whom do you need to get to know better?
- How much knowledge do you have of these people and how much do they know about you?
- What gaps are there in your company's network and how is it planned to fill them?
- What is the policy/action plan regarding corporate events/entertaining?
- Is there a regular review meeting for business-relationship development?
- Develop your rapport-building web/establish targets and measure results regularly and often.
- Have any recent customer-satisfaction surveys been carried out?
- If so, what information did they yield and what use was made of it?

10.4.5.2 Vary your approach

There are certain ways to maximize success and continue development. Vary your approach to relationship building: here are five ways.

1. **Business network connections**: By designing and building a unique business-network system you can track the effectiveness of contacts, collaborators, strategic alliances, complementary firms, products and services. They can add value in so many ways.
2. **Connecting individuals**: By harnessing the power of personal connections you can utilize to your best advantage networks, connections and relationships. Individuals can offer tailored introductions to you to help you with your career progression or business development.
3. **Connecting with customers**: If you use your individual connections wisely you can develop connection strategies to win, retain and develop new business. By means of regular audit of existing and past customer relationships, you can win new projects and open up new markets.

4. **Connecting with staff**: Be passionate about relationship building within your company and empowering individuals and teams. It is possible to deliver outstanding results by means of programmes designed to encourage connections between employees and employers.
5. **Discreet connections**: If you have sensitive business connections – for instance, in areas concerning potential sales and acquisitions of a business division, investment procurement or executive searches and headhunting projects – by using your unique professional network, solutions are often found speedily and successfully.

10.4.6 Perseverance (and more) will get results

To harness the power of personal connections you need to keep a few 'P' words in mind.

- **Persistence** pays, there's no doubt about that.
- Relationship building is like **planting seeds** – they take time to germinate.
- One of the most important factors in the process is **preparation**.
- You need to prepare the ground – it pays to know about your **prospects**.
- You have to **persevere** – sometimes for months, and in some cases years.
- Try to be unfailingly **polite** and **patient**.
- A **positive** mental attitude and outlook is infectious.
- **Persuasion** tactics get easier with **practice**.
- Make sure you do some **planning** – it helps you to know when and how to keep in contact.
- Don't underestimate the value of **praise** when communicating with your business contacts.
- Most people respond **positively** to flattery.

10.4.7 Say it with feeling

What's important is the way you begin to build business relationships. In effect you're starting a process of persuasion. It's not easy – often just using words is not enough. To be persuasive you should offer people reasons that reflect their point of view.

> **Networking hint**
> You have to be able to persuade the other party into the idea that there is something in it for them.

Benefits are things that do something for people. The benefits of reading this book include helping you with your strategy for building valuable business relationships.

10.4.8 Touching emotions and intellect

To be an effective persuader you should not only offer good reasons for something but also create emotional goodwill at the same time. If you need to persuade powerfully, bring in stories to connect with people's hearts as well as minds.

10.4.9 Recommendations work wonders

If you have shared connections, it is much easier to persuade when there are credible people to testify that your skills helped them in some way or other.

If possible, don't rely on one source for recommendations. Using several different parties gives further weight to your case. You increase your chances that one or other of your sources will be a powerful influence over the person with whom you're building up trust.

10.4.10 Don't waste their time

- Think about how you want to come across to your business contact.
- Ask yourself why anyone should want to listen to you.
- List your reasons and then organize them.
- What are the most important things you are trying to say.
- How can you build rapport with one another?
- Can you arrange your thoughts into a logical sequence?
- You could start with something attention grabbing and continue to maintain interest throughout the exchange.
- Perhaps you want to build up your case throughout the dialogue and end with some weighty fact that has masses of impact.

10.4.11 Check progress

By using a simple achievement matrix you can check progress and remind yourself of your continual plan for refreshing and reviving your brilliant business connections.

Step 1

- Review your contacts database and list your achievements and successes over the last month.
- How many new contacts have you added?
- How many existing contacts did you manage to reach? Were these by telephone, email or face-to-face meetings?

Step 2

- List the obstacles that you have overcome and those that are yet to be solved.
- Was it a time-management issue?
- Did you attend a sufficient number of networking events?
- Did you develop the appropriate method of communication?

Step 3

- Make a note of the action you will need to take to resolve the latest obstacles.
- Do you need to phone more people?
- Have you progressed certain connections as far as possible?
- Are you awaiting information from other people that they promised you?

168 *How to build successful business relationships*

- Have you followed through with the information, introductions, etc, that you have offered people in your network?

Step 4

- List the objectives for the next period.
- How many people do you wish to connect with by the end of the month?
- What percentage of these are existing contacts and how many are new introductions?
- Are the work-getting targets progressing according to plan?

10.4.12 Communication skills awareness checklist

- **Presence**: Pay attention to the way your voice and body language are used in conjunction with the words you speak. You can convey the right impression if they are used correctly.
- **Relating**: Don't underestimate the importance of developing your rapport-building skills to get on the same wavelength of your business prospect.
- **Questioning**: When engaged in conversation with your contact, make sure you match your question to the situation or subject. Beware asking irrelevant questions: this will show that you've not paid attention to what she said.
- **Listening**: Listen to everything your contact says attentively. Try to reach 'Level 4: attentive listening' (see Chapter 6, specifically section 6.5, for the five levels of listening skills). If she's likely to become a significant influence in your business development strategy you should aim for 'Level Five: empathy' eventually.
- **Checking**: This is the art of glancing at your business contact to see that she's still on your wavelength while you're engaged in dialogue. Watch for gestures and see whether she does the same when she's talking to you.

10.4.13 Checklist for developing powerful relationships

- Be transparent in your actions.
- Communicate with all sides as well as upwards and downwards.
- Network extensively to keep well informed.
- Identify and watch the 'politicians'.
- Put yourself in other people's shoes.
- Anticipate and manage others' reactions.
- Be clearly good at your job.

10.4.14 Every relationship is different

Without business contacts, preferably positive ones, you will make little progress. In essence the key to building brilliant business connections is being sincere:

- establish **trust** – respect your business contacts and your staff; they will respond in a positive way;

- remember to say **thank you** – it costs you nothing and gains you much;
- **treasure the moment** if someone says, 'Thank goodness we met!';
- you may get **testimonials** from satisfied clients or recommenders;
- when other people **blow your trumpet** for you the sound is much louder;
- you will find your thinking processes have changed: you will be far **more people-oriented** – whether externally or internally;
- With **tenacity** these relationships will **flourish**.

> **Networking hint**
> Your contacts are unique, some are demanding, others quite difficult, and some require understanding. Understanding them doesn't just make your relationship building easier: it is inherent to the whole process. Where there is no understanding, how can a business relationship flourish?

10.5 And finally . . .

If you harness the power of personal connections, you will realize the importance of five words:

- enquire
- listen
- offer
- trust
- respect

This book should help you recognize the power that relationship building can have on business and people. It can make things easier and more effective. Remember, it's *people*, not just skills, education, qualifications and experience, that matter. People do business with people they like and trust.

And, when it comes to developing valuable business connections, remember this:

Small talk really means big business.

Index

Adding value 146
Anger at work 29–30
Appraisals, performance 109–12
Appointing staff 114–19
Appraisals 109–12
Approaches
  new 38
  systematic 35
Assertive behaviour 26–8, 97–8
Attention, paying 100
Attitudes and approaches 34, 54, 69–75
Attributes, positive and negative 7
Awareness, of others 12

Behaviour 7
  assertive 27–8
  difficult 56–64
  negative 7
  positive 7
Benefits, importance of 146
Best practice skills 157
Body language 81–2
Building relationships 44–51
Business connections 33–9
Business introductions 16
Business relationships 1

Character types 54–7, 80
Charm 42, 76
Checklists for success 159–69
Client relationships 17
Colleagues 2, 53–64
Coincidences 37
Communication
  dealing with others 10
  difficulties, overcoming 28
  directing the cycle 90
  forms of 91–3
  getting started 85

non-verbal 94
opening rituals 91
skills checklist 93, 167
successful 10
Competitor analysis, value of 154–6
Confidence 71, 96–101
Connecting, suggested methods 33–51
Connections, categorizing 36, 43
Contacts, use of 164
Conversational techniques 25, 90–2
Courtesy, effects of 140
Credible, being 146
Criticism, dealing with 30–1
Curiosity 140
Customer satisfaction survey 150–4
CVs 117–19

Database 33–7, 43
Decisions, effective 9
Delegation techniques 103–7
Difficult people 58
Disciplining staff 112–14
Distinctive, being 44
Dynamics, group and team 123–38

Ears, use of 89–90
Effective, being 6
Email 93
Emotions, feelings 166
Empathy, creating 140
Empowerment 124, 128–37
Encounters, face to face 88–9
Enquiries 140
Envy, professional 97
Eye contact 99

Face to face encounters 40–1, 88–9
Feedback 6
Feelings, emotions 166

First impressions 2, 73
Flexibility 12

Good manners 100
Group dynamics 123–38

Herzberg's law 130

Identity, brand and company 142
Impact, creating 3
Impressions
 first 2, 73
 of others 10
 visual 99
Influential people 78
Interest, creating 140
Involvement 124

Jargon, use of 141
Jealousy, professional 97

Keeping in touch 43
Key players 78
Knowledge 71

Letters 93
Links, importance of 42–4
Listening skills 89–90

Making connections 33
 seven steps for 42–4
Managing others 103–19
Manners, mannerisms 94
Meetings 40–1, 64–8
Mentoring 160–2
Morale 95
Motivation 125–38
 fundamentals of 128–37
 Herzberg's law 130
 successful 126, 132, 138
 Theory X and Y 130
Movers and shakers 79–80

Networking
 importance of 18
 strategy 69–71, 83
Non-verbal communication 94
Notes, keeping 43

Objections, handling 11
Office politics 60
Office who's who 33–5
Opportunities, creating 83

Pareto's Law 143
Partnering 156–8
People management 103–19
Performance appraisals 109–12
Perseverance, importance of 50, 166–7
Personal attributes 7
Personal touch 148
Personality types 34
Persuasive, being 50–1, 145–9, 166
Politics, office 60
Positive behaviour 71
Powerful relationships 168
Praise 95, 140
Prioritizing 5, 72
Professional relationships 154
 development of 154–8
Progress, check 167–70

Questions
 the art of good 42, 140

RAPPORT mnemonic 18
Recommendations 142–4, 164–6
Reconnaissance 74
Recruitment 114–19
Referrals 139–43
Relationships, business
 building 1, 15–18, 44–51
 checklist 159–69
 different types of 144–58
 rewards of 139
 results of 139
Respect, showing 141
Responsibility, power of 124
Reviewing progress 35
Risk awareness 157

Self-management 1–6
Self-sufficiency 123
Sincerity, importance of 141
SMART working 4, 47
Succession planning 17
Staff
 appraisals 111

delegation techniques 104
disciplining 113
interviews 110
jealousy 97
performance 109
potential 150
recognition of good service 150
recruitment 115
selection 115
valuing 149
Strategic alliances 16, 156–8
Success, checklists for 159–69

Teams
building 17
meeting the team 121
motivating the team 125–38
strengthening the team 124
team dynamics 123
working together 122
Telephone calls 92
Text messages 92
Theory X and Y 130
Thinking, applied 36
Three-stage plan 17

Virtual team, creating 83
Vocal skills 85–8
Voice, use of 85–8
Voicemail messages 92

Winning friends 76
Workplace relationships, internal 53, 60
Written communication 93